T0253396

# LIVING WITH URBAN ENVIRONMENTAL HEALTH RISKS

To the memory of
Belaynesh Gebre Michael–my beloved mother


*All that I am or ever hope to be, I owe to my angel mother*
Abraham Lincoln

# Living With Urban Environmental Health Risks

## The Case of Ethiopia

GIRMA KEBBEDE
*Mount Holyoke College, USA*

Routledge
Taylor & Francis Group

LONDON AND NEW YORK

First published 2004 by Ashgate Publishing

2 Park Square, Milton Park, Abingdon, Oxfordshire OX14 4RN
52 Vanderbilt Avenue, New York, NY 10017

*Routledge is an imprint of the Taylor & Francis Group, an informa business*

First issued in paperback 2019

Copyright © Girma Kebbede 2004

Girma Kebbede has asserted his right under the Copyright, Designs and Patents Act, 1988, to be identified as the author of this work.

All rights reserved. No part of this book may be reprinted or reproduced or utilised in any form or by any electronic, mechanical, or other means, now known or hereafter invented, including photocopying and recording, or in any information storage or retrieval system, without permission in writing from the publishers.

Notice:
Product or corporate names may be trademarks or registered trademarks, and are used only for identification and explanation without intent to infringe.

A Library of Congress record exists under LC control number: 2004009028

Publisher's Note
The publisher has gone to great lengths to ensure the quality of this reprint but points out that some imperfections in the original copies may be apparent.

Disclaimer
The publisher has made every effort to trace copyright holders and welcomes correspondence from those they have been unable to contact.

ISBN 13: 978-0-8153-9031-2 (hbk)
ISBN 13: 978-1-138-35834-8 (pbk)

# Contents

*List of Tables*                                                    *vi*

*Preface and Acknowledgements*                                      *vii*

*List of Abbreviations*                                             *x*

*Ethiopia: Regional States and Major Urban Centers*                *xiii*

1.  Introduction: The State and Development                         1

2.  Urban Growth and Decay                                          22

3.  Urban Water Supply                                              74

4.  Urban Sanitation and Waste Management                           107

5.  Pollution                                                       145

6.  The State of Health and Health Services                         171

7.  Summary and Conclusion                                          218

*Bibliography*                                                      *231*

*Index*                                                             *247*

# List of Tables

| | | |
|---|---|---|
| 2.1 | Population Growth of Urban Centers with Population 50,000 and Above, 1967 – 2001 | 33 |
| 2.2 | Informal Settlements in Addis Ababa: Location and Size | 52 |
| 2.3 | Major Ongoing Projects in Addis Ababa: Real Estate and Other Infrastructure | 56 |
| 3.1 | Sources of Drinking Water by Place of Residence, 1998 | 91 |
| 3.2 | Distribution of Households by Sources of Drinking Water in Selected Urban Centers, 1998 | 92 |
| 3.3 | Common Diseases Transmitted to Humans Through Contaminated Water | 104 |
| 4.1 | Distribution of Households by Toilet Facility in Selected Urban Centers, 1998 | 108 |
| 4.2 | Distribution of Households by Method of Solid Waste Disposal in Selected Urban Centers, 1998 | 123 |
| 4.3 | Categories of Healthcare Waste | 128 |
| 4.4 | Examples of Infections by Exposure to Healthcare Wastes, Causative Organisms and Transmission Vehicles | 129 |
| 4.5 | Environmental Classification of Excreta-related Diseases | 140 |
| 4.6 | Control of Excreta-related Diseases | 142 |
| 5.1 | Distribution of Manufacturing Industries by Regional States | 147 |
| 5.2 | Methods of Wastewater and Solid Waste Disposal for Selected Industries in Addis Ababa | 150 |
| 5.3 | Gaseous Air Pollutants | 160 |
| 5.4 | Types of Cooking Fuel Used by Households in Selected Urban Centers, 1998 | 163 |
| 6.1a | Development of Healthcare Facilities, 1947/48 – 2000/2001 | 173 |
| 6.1b | Development of Health Human Resources, 1947/48 – 2000/2001 | 173 |
| 6.2 | Healthcare Facilities by Regional States, 2000/2001 | 182 |
| 6.3 | Maternal Health Service Achievement and Coverage by Regional States, 2000/2001 | 187 |
| 6.4 | Immunization Coverage by Regional States, 2000/2001 | 189 |
| 6.5a | Top 10 Leading Causes of Outpatient Visits, 2001 | 204 |
| 6.5b | Top 10 Leading Causes of Hospital Admissions, 2001 | 204 |
| 6.5c | Top 10 Leading Causes of Deaths, 2001 | 205 |

# Preface and Acknowledgements

Most urban inhabitants in Ethiopia live in an impoverished environment in extremely squalid and overcrowded housing and neighborhoods. In Addis Ababa, the capital, most houses are utterly unfit for human habitation. The sanitation situation in the city is appalling. People defecate by rivers and streams, at roadsides, and near public buildings for lack of adequate public sanitation facilities. Heaps of stinking, fly- and rodent-infested wastes appear everywhere, blocking neighborhood streets and storm drainages. Municipal bins overflow with garbage, and people dump waste on street corners and in storm drains. Light rains cause streets to flood because uncollected waste blocks drainage channels. Children and young adults play next to waste dumps for lack of dedicated recreational areas and open spaces. The young and the old alike go about their daily business oblivious of the filth around them. The municipal government lacks the capacity and competence to improve the situation.

Largely unregulated and heavily concentrated in Addis Ababa, industries release harmful pollutants into the air, water, and open spaces, compromising the health of urban residents and surrounding rural inhabitants and their means of livelihoods. The federal government has policies to deal with urban environmental pollution, but very little progress has been achieved owing largely to failure to translate polices into practice. A public Environmental Protection Authority does exist but it is not a well-staffed regulatory agency with the ability to enforce existing laws, nor does it have the power to impose penalties on offenders breaking environmental laws.

Considering the major components of the built environment such as the quality of housing stock and basic physical infrastructure—including water, sanitation, drainage, roads and public transportation networks, parks and recreational facilities, and other public amenities—urban areas in Ethiopia are far from supporting a healthy life. Because of poor environmental conditions and the lack of adequate urban public services necessary for good health, diseases such as diarrhea, pneumonia, tuberculosis, and acute respiratory infections remain major causes of mortality, especially of children.

The main objective of this study is to examine the state of the urban environment in Ethiopia. The study methodology combines field investigations in a number of major urban areas in the country and the collection and synthesis of existing literature on urban areas, including published and unpublished government reports, censuses, and sample surveys; proceedings of numerous conventions and workshops organized by government institutions, non-governmental organizations, and professional associations; and expert reports and articles published in local academic journals, newspapers, and magazines.

Field investigations were done in Addis Ababa and other major urban areas, including Adama, Awasa, Bahir Dar, Bishoftu, Dire Dawa, Gondar, Harar, Jimma, and Mekele. The principal methods of field inquiry included discussions with pertinent municipal and healthcare personnel and non-governmental organizations, individual and group interviews with residents, and inspections of water, sanitation, and healthcare facilities.

This study was made possible through the generous financial support of two institutions: Mount Holyoke College, my home institution, provided the sabbatical leave and faculty grants to carry out a series of fieldwork, while a 2001-2002 fellowship award from the Fulbright Program allowed me to conduct further fieldwork and to complete a draft that has culminated in this volume. I am, therefore, deeply grateful to both institutions.

I am indebted to many friends in Ethiopia who have directly or indirectly helped me in collecting information for this book. To Engidashet Bunare, who provided invaluable insights, encouragement, and assistance in gathering information on water and sanitation services in urban areas, I express my deep and permanent gratitude. I owe special thanks to Girum Tadesse and Marilyn Sherman, both health professionals. Girum assisted in gathering information on health services and read through the first draft of the chapter on health and made constructive criticisms about facts and interpretation of data. Through talking to him I learned a great deal about healthcare issues in Ethiopia. Marilyn also read the health chapter and offered useful ideas suggestions. I greatly appreciate Sisay Tefera's valuable advice and help in translating Amharic documents and in conducting health-related surveys in Addis Ababa. I also wish to thank Yigzaw Ayalew, a researcher at the Environmental Protection Authority, for allowing me to access conference proceedings and workshop papers in his possession and for guiding me to the right sources of information.

I am also grateful to many government officials—who prefer to remain anonymous—for allowing me to access public documents. Without their cooperation, it would have been difficult, if not impossible, to complete this book. I would like to thank, too, Alemaya University for providing logistical support for fieldwork. I am also deeply grateful to the family of Tegegne Sishaw whose unusual kindness and hospitality made my stay at Alemaya University as a Fulbright scholar enjoyable and productive. My profound gratitude also goes to my long-time friends, Professor Ayele Bekerie, who provided detailed comments and suggestions that helped to improve various sections of the manuscript, and to Yebza Lakew, for his useful comments on governance and decentralization issues discussed in Chapter 1.

There are many others who have contributed to this book with kindness, information, and suggestions, and it would be impossible to list them all. I therefore reiterate my heartfelt gratitude to all who have helped me in different ways and degrees. While I gratefully acknowledge all the support I received, the findings, interpretations, and conclusions expressed in this book are mine and are not necessarily shared by those who kindly provided all sorts of help. The responsibility for any factual errors that appear in this book is also mine alone.

Last, but certainly not least, much credit goes to my wife Aster and daughters Sophia and Hanna for their love and support and for patiently tolerating the prolonged absences from home that this research involved.

Girma Kebbede
South Hadley, Massachusetts

# List of Abbreviations

| | |
|---|---|
| AA | Addis Ababa |
| AACG | Addis Ababa City Government |
| AAMPRO | Addis Ababa Master Plan Revision Office |
| AARH | Agency for the Administration of Rental Houses |
| AAU | Addis Ababa University |
| AAWSA | Addis Ababa Water and Sewerage Authority |
| ACD | Acute Childhood Diarrhea |
| ADLI | Agriculture Development Led Industrialization |
| AIDS | Acquired Immune Deficiency Syndrome |
| BG | Benishangul-Gumuz |
| BOD | Biochemical Oxygen Demand |
| CAFOD | Catholic Agency for Overseas Development (UK) |
| CBB | Construction and Business Bank |
| CBO | Community-based organization |
| CFCs | Chlorofluorocarbons |
| CIDA | Canada International Development Agency |
| COD | Chemical Oxygen Demand |
| CSA | Central Statistical Authority |
| DO | Dissolved Oxygen |
| DPT | Diphtheria, Pertusis, and Titanus |
| ECA | United Nations Economic Commission for Africa |
| EPA | Environmental Protection Authority/Ethiopia |
| EPI | Expanded Program on Immunization |
| EPRDF | Ethiopian People's Revolutionary Democratic Front |
| ESA | Ethiopian Standardization Authority |
| EVDSA | Ethiopian Valleys Development Studies Authority |
| EWLA | Ethiopian Women Lawyers' Association |
| EWWCA | Ethiopian Water Works Construction Authority |
| EWWCE | Ethiopian Water Works Construction Enterprise |
| FDRE | Federal Democratic Republic of Ethiopia |
| FGAE | Family Guidance Association of Ethiopia |
| FGM | Female Genital Mutilation |
| FHI | Family Health International |
| GDP | Gross Domestic Product |
| GNP | Gross National Product |
| GTZ | Deutsche Gesellschaft Fur Technische Zusammenarbeit (German Agency for Technical Cooperation) |

| | |
|---|---|
| HIPCs | Heavily Indebted Poor Countries |
| HIV | Human Immunodeficiency Virus |
| HSB | Housing and Savings Bank |
| HTPs | Harmful Traditional Practices |
| IMF | International Monetary Fund |
| ILO | International Labor Organization |
| ISAPSO | Integrated Service for AIDS Prevention and Support Organization |
| LDPE | Low-density polyethylene |
| MDG | Millennium Development Goals (UN) |
| MEDaC | Ministry of Economic Development and Cooperation |
| MIDROC | Mohammed International Development, Research, and Organization Companies |
| MNRDEP | Ministry of Natural Resources Development and Environmental Protection |
| MOH | Ministry of Health |
| MOLSA | Ministry of Labor and Social Affairs |
| MOWR | Ministry of Water Resources |
| MPWUDH | Ministry of Public Works and Urban Development and Housing |
| MUDH | Ministry of Urban Development and Housing |
| MWUD | Ministry of Works and Urban Development |
| NCTPE | National Committee on Traditional Practices in Ethiopia |
| NMSA | National Meteorological Service Agency |
| NWRC | National Water Resources Commission |
| OSSA | Organization for Social Services for AIDS |
| PHC | Primary Health Care |
| SNNPR | Southern Nations, Nationalities, and Peoples' Region |
| SPM | Suspended Particulate Matters |
| STDs | Sexually Transmitted Diseases |
| STI | Sexually Transmitted Infection |
| TB | Tuberculosis |
| TGE | Transitional Government of Ethiopia |
| TPLF | Tigray People's Liberation Front |
| UDAs | Urban Dwellers Associations |
| UMAS | Urban Management Advisory Services |
| UNAIDS | United Nations Program on HIV/AIDS |
| UNCED | United Nations Conference on Environment and Development |
| UNCHS | United Nations Center for Human Settlements (Habitat) |
| UNDP | United Nations Development Program |
| UNECA | United Nations Economic Commission for Africa |

| | |
|---|---|
| UNEP | United Nations Environment Program |
| UNESCO | United Nations Educational, Scientific, and Cultural Organization |
| UNICEF | United Nations (International) Children's (Emergency) Fund |
| UNIDO | United Nations Industrial Development Organization |
| UNOCHA | United Nations Office for the Coordination of Humanitarian Affairs |
| USAID | United States Agency for International Development |
| VOC | Volatile organic compounds |
| WHO | World Health Organization |
| WMO | World Metrological Organization |
| WRDA | Water Resources Development Authority |
| WRI | World Resources Institute |
| WSSA | Water Supply and Sewerage Authority |

# Ethiopia: Regional States and Major Urban Centers

# Chapter 1

# Introduction: The State and Development

The environment in which people reside has an enormous influence on their quality of health. Microorganisms—viruses, bacteria, protozoa, and helminths—are the most dreadful environmental threat to humans. Most infectious diseases are associated with conditions in the physical environment (World Bank, 1998: 18). Ill health abounds in Ethiopia because of the pitiable environmental conditions such as poor shelter, overcrowding, unsafe drinking water, poor sanitation, water pollution, and indoor and outdoor air pollution. It has been estimated that water borne pathogens responsible for cholera, typhoid, amoebic infections, bacillary dysentery, and diarrhea account for four-fifths of all disease in the country. These environmental conditions are, in return, intimately related to poverty and the general lack of economic and political progress in the country. The poor are especially exposed to the worst environmental health risks.[1]

*Living With Urban Environmental Health Risks* examines the extent and nature of environmental problems in urban areas in Ethiopia and their impact on health. Although Ethiopia is one of the lowest urbanizing countries in the world, with about 80 percent of its people still living in the rural areas, it has been experiencing one of the most rapid urbanization processes in recent years. Between 1984 and 2000, urban areas with a population of 50,000 or more grew at an average rate of 6.7 percent per year. Such a high growth rate, however, was not accompanied by a commensurate increase in the basic urban amenities that are indispensable to an adequate and healthy urban environment. Urban governments in the country are unable to provide or facilitate investment in the necessary housing, infrastructure, and amenities to accommodate urban growth. The urban environment, particularly that of the capital, Addis Ababa, is under constant stress due to the pressures of poor environmental infrastructure, rapid population growth, and inadequate planning controls. The majority of the urban population is badly housed and lacks basic services. Housing, water supply, sanitation services, drainage, transport network, and health services have not been able to keep pace with the prevailing urban growth rates. This has resulted in deteriorating urban living conditions and increasingly serious health problems. This book argues that these problems are largely the result of ineffective urban governance and the national government's wrong land use policies that work against the interests of urban citizens and their natural and built environment.

**The Focus on Urban Areas**

Urban areas in Ethiopia will continue to grow not only due to the natural increase of their own populations but also due to continued rural-urban migration. While Addis Ababa will remain the most attractive destination for rural and small town migrants, the decentralization policy of the government has stimulated the growth of second-tier urban areas, especially the regional capitals. Many second-tier urban areas (for example, Adama, Mekele, Bahir Dar, and Awasa) are growing much faster than Addis Ababa. Population forecasts indicate that over the next 20 years the country's population will double. That means by the year 2025 an additional 60-65 million people will need environmental health services. Many of these services will focus on the country's rapidly growing cities and towns. By the year 2025, the country will have 20 or more cities with more than 100,000 inhabitants and 30 or more cities with populations ranging between 50,000 and 100,000. The challenge to provide essential urban amenities to the residents of the cities and smaller urban areas will be overwhelming.

Although fewer than 20 percent of Ethiopians live and work in urban areas, cities and towns account for a considerable proportion of the country's total economic activity. Urban areas represent a huge 'investment in terms of capital and human resources' (Jowsey and Kellett, 1995: 197). More than one half of the country's gross domestic product is generated in urban areas (World Bank, 2003: 238). An estimated 90 percent of the industrial activity is urban based. It is in urban areas that a disproportionate amount of resources are used and wastes detrimental to human health and the environment are produced. In spite of all this, not much attention has been given to urban areas. There exists no national policy that guides urban development. Urban areas pursue haphazard development without any guidance or systematic planning and integration to their rural hinterlands.

Urban areas, especially large ones, are exposed to health threats uncommon in rural settings. Pollution of water and air as a result of industrial and transportation activities is a prime example. The incidence of infectious diseases (for example, acute respiratory infections, tuberculosis, and other airborne infections) tends to be more prevalent in crowded urban areas than in rural environments. Urban environments, particularly poor neighborhoods with poor sanitation, water, and solid waste services, are hosts to vermin that transmit diseases. Inappropriate disposal practices of municipal solid waste and industrial and healthcare wastes imperil public health. For the most part, Ethiopian urban areas display wide ranging problems, including rapid population growth, unemployment and underemployment, inadequate shelter, poor healthcare and sanitation facilities, poorly maintained physical infrastructure, and environmental pollution.

The focus on urban environmental problems by no means suggests that rural areas suffer from fewer environmental problems, nor does it suggest that rural areas have access to adequate healthcare and essential social and economic infrastructure. A number of major studies have already documented the problems of rural Ethiopia. By comparison, few address the problems of urban areas. Nor does the focus of the book implicitly advocate an increase in rural-to-urban resource transfer. Urban areas already possess copious human and physical resources; however, poor management and misguided land use practices and development policies have squandered these resources.

Moreover, the findings of this book may have implications for the welfare of the rural population as well. Improvements in the urban environment and in the livelihood of its people will have a positive impact on the surrounding rural areas. River pollution, for instance, can contaminate downstream environments far from its sources. In this case, the proper management of solid wastes, the treatment of wastewater, and the control of pollution will therefore improve the condition of rivers, lakes, and soils that sustain the immediate rural areas. In addition, a clean urban environment will ensure a healthier population and thus incur less medical expenditures. These extra resources could then be allocated to rural areas. Improving urban livelihoods could also contribute immensely to rural development. Ideally, urban areas provide a market for rural products, produce and enhance industrial goods and services for their hinterlands, and provide non-farm employments for rural surplus labor (World Bank, 2003: 107). At present the role of urban centers as growth poles in Ethiopia remains extremely limited.

This chapter seeks to set the urban environmental problems identified and discussed in this volume within the broader context of the persistent political and economic development dilemmas that the country continues to endure. The chapter contends that the lack of genuine economic and political development in the country stems not from insufficient physical and human resources, but from unaccountable governments and development policies that have failed to respond to the needs and priorities of the people living both in the urban and the rural areas.

## The Burdens of Misrule

Ethiopia has considerable resources; nevertheless, it lingers at the tail end of the social, economic, and political development continuum. One of the five or so poorest countries in the world, Ethiopia has an estimated income of just $100 a year per person (World Bank, 2003: 14).[2] The vast majority of the population lives in acute poverty and suffers from inadequate access to safe drinking water, sanitation facilities, decent shelter, healthcare, and education. Life expectancy is less than 52 years and is expected to decrease by the impact of the AIDS pandemic. Most preventable diseases go untreated due to scarce medical services. The annual income of the average household cannot adequately sustain the life of a single soul.[3] Chronic malnutrition remains a crucial problem. The country already has basic resources, and in fact has the capacity to meet the essential necessities of its citizens. Unfortunately, throughout its modern history, Ethiopia has lacked accountable and visionary political leadership necessary to get the most out of its resources and people. Bad governance has hindered the country's development.

During the imperial regime that ended in 1974, the country experienced an extremely exploitative socioeconomic system and an oppressive political order. Unequal distribution of the country's means of production—land— deprived peasant farmers, who accounted for over four-fifths of the country's population, of the opportunity to improve their lives. The landed aristocratic rule marginalized the agrarian sector by expropriating agricultural surpluses in the form of land rent and other types of mandatory payments (Kebbede, 1992: 1). It exacted revenues from the farming population that far exceeded their actual capacity to pay. Extremely low wages and strict prohibition against the creation of autonomous trade unions kept the urban working class in an appalling state of material conditions. Recurrent hunger and devastating famines took the lives of many of the most productive forces of the country. Overall, the pre-1974 political and economic system kept the vast majority of the population trapped in poverty, ignorance, and disease.

From 1974 to 1991, Ethiopia experienced a degree of misfortune unprecedented in the modern history of the country. From the mid-1970s through the 1980s, the country experienced a pitiless regime of centralized control. All rural land became state property; farmers did not have ownership rights over the land they cultivated. The government forced farmers to collectivize and determined what crops would be grown and at what price they would be sold. Because they could not own or control either the land they farmed or the crops they yielded, farmers did not use their energy, time, or knowledge to improve their productivity and better their lives. Forced resettlement and villagization schemes wrecked the social, cultural and economic life of the rural sector. The regime's state-mandated development strategy and anti-private sector policies strangled growth and development in the urban sectors of the economy. The commercial, industrial, and urban housing sectors suffered immensely from bureaucratic control and management and restrictive state measures against private investments. The economy of the country stagnated throughout much of the 1970s and 1980s. Famine and secessionist wars took the lives of millions. Political viciousness annihilated a whole generation of young people and forced hundreds of thousands of the country's educated and skilled people into a life of exile (Kebbede, 1992).

The year 1991 presented a turning point for Ethiopia. The notorious Derg regime of Colonel Mengistu Haile Mariam collapsed like a house of cards. The Ethiopian People's Revolutionary Democratic Front (EPRDF)—largely formed from the Tigray People's Liberation Front (TPLF)—came into power in the wake of a long-drawn-out and destructive civil war. In 1993, the new government agreed to the independence of Eritrea, land-locking Ethiopia. The EPRDF argued that only in this way could the remnant of Ethiopia stay intact.

The new rulers restructured the internal administration of the country from highly centralized control to decentralized autonomous regional states created largely on the basis of linguistic identities.[4] Nine linguistic regions make up the Ethiopian Federal Union: Tigray, Afar, Amhara, Oromiya, Somali, Benishangul-Gumuz, Southern Nations, Nationalities and People's Region, Gambela, and Harari. The largest two cities of the country—Addis Ababa and Dire Dawa—are non-autonomous administrative and council states, respectively. Each of the nine regional states is, in turn, subdivided into zones, *weredas*, and *kebeles*. The country has in total 62 zones, 534 *weredas* or districts, and about 10,000 *kebeles* or neighborhoods. Under the new decentralized state structure, ethnic regions are allowed—even encouraged—to maintain their cultural identities and languages. Article 39 of the Federal Constitution also recognizes the fundamental right of self-determination for regional states up to and including secession.

In 1995, Ethiopia established a federal system of government with a two-tier parliament: the House of Federation or upper chamber with 117 seats (chosen by state assemblies, members serve five-year terms) and the House of People's Representatives or lower chamber with 548 seats (directly elected by popular vote from single-member districts, members serve five-year terms). The House of People's Representatives elects the President of the Federal Republic for a six-year term of ceremonial function while the party in power following legislative elections designates the Prime Minister to serve a five-year term.

The federal constitution accords regional states a number of rights. The regional state may restructure its internal jurisdictions (zones and *weredas*) on the basis of ethnic and linguistic interests where it deems necessary, establish a self-governing constitution, develop social and economic policies and strategies, administer its natural resources in accordance with the federal laws, assess and collect taxes, and establish security forces to maintain peace and order. The responsibilities of the federal government include national security (defense), foreign relations, the granting of citizenship, economic policies, the declaration of state of emergency, currency, and the establishment and overseeing of key development programs and communication networks (Gebre Egziabher, 1998: 695).

Regional states ought to have fiscal independence by generating their own revenues through taxes and fees. In reality, however, they do not. First, the unequal distribution of resources among the regions renders some states less able than others to raise adequate revenues to meet their bureaucratic and development needs. For instance, regional states such as Tigray, Afar, Somali, Gambela, and Benishangul-Gumuz command much fewer resources than the Amhara, Oromiya, and the state of Southern Nations, Nationalities and Peoples. The latter, by virtue of their physical size and resource endowment, have a strong tax base—even though they also serve a larger population. Second, many regions lack institutional capacities to develop their revenue base and efficiently collect taxes. Finally, the federal government limits the tax sources of the states (Gebre Egziabher, 1998: 695). Hence, most of the newly constituted autonomous regional states, including the resource-rich ones, suffer from a weak financial position and thus depend on the revenue of the federal government to pay salaries and to build and maintain social and economic infrastructure. The fiscal over-dependence of the regional states has a number of drawbacks that may cast doubts on the authenticity of the decentralization effort. First, because locally raised revenues constitute only a small portion of their budgets, regional governments may not have adequate fiscal control to plan their development activities. Second, too much financial dependence on the center can weaken locally designed development policies and programs. Third, regional authorities may not have the incentive to put federal funds to proper and efficient use.

The constitution grants regional states the power to form their own legislative, executive, and judiciary branches of government—mirroring the federal government's administrative structure. As a result, regional states have huge resource-consuming administrative apparatuses. After paying administrative salaries and wages, regional states find it difficult to fund development programs. Regional state administrative bureaucracies have expanded while the revenues needed to support them have not. Additionally, in many regional states—with the possible exception of Tigray—cumbersome bureaucracies and corrupt state officials thwart local and outside initiatives that seek investment opportunities. Local and regional state authorities cannot provide effective responses to local problems. Moreover, unless a hefty kickback lures them, they often place stumbling blocks against private initiatives.

In regional states, security forces (with federal security forces, in some instances) and local authorities routinely violate citizens' rights at will, without any fear of legal repercussions. Some state and local political and security authorities seem to hold the view that the law does not apply to officials. The legal and judicial system that supposedly upholds the constitutional rights of individuals is totally corrupt. The judiciary is not independent of the legislative and executive branches. The government-owned mass media often reports lawlessness and violence against poor citizens, especially women, in regional states, however, the federal government cannot monitor the compliance of regional institutions with the law, nor can it safeguard against abuse of power in regional states. At present, the laws of the country are not adhered to outside the capital city.

## Democratic Pretension

Upon assuming power in 1991 the ruling party promised to facilitate the evolution of democratic political institutions in the country. It pledged to install good governance that would promote the principles of democracy, accountability, and equality. It also vowed to dismantle the state-mandated development strategy pursued by the previous regime in favor of a private economy to hasten the country's revitalization. However, many political observers argue that the ruling party has failed to live up to its grand promise. Nearly twelve years after the EPRDF assumed power, genuine democracy has yet to take root in Ethiopia.[5] Parliamentary elections have taken place twice in the country; however, critics say that neither election was conducted in a manner that was free and fair for all contestants. The ruling party designed an electoral system that made it difficult for opposition political groups to mobilize their supporters and to secure financial resources behind their candidates. Opposition candidates were harassed and denied access to the mass media in order to put their ideas to the electorate. Supporters of opposition candidates were beleaguered for freely expressing their will in the ballot box.[6] On the other hand, candidates affiliated with the party in power received adequate financial resources and media access to mobilize support. In nearly all national and regional contests, candidates of the ruling party ran virtually unopposed. In both elections, everyone knew the election results before the voting took place. The candidates of the ruling party—the EPRDF— captured nearly all the parliamentary seats. In its present form, opposition parties argue, the parliament is nothing but a rubber-stamping institution. They say legislations drafted by the executive branch become laws with few or no changes in parliament. The views of opposing parties are disparaged and dismissed. Because affiliates of the EPRDF represent the majority in all regional states, a diversity of policy decisions is lacking there, too.

Although the elections have given the ruling party the appearance of legitimacy, they have not upheld the democratic rights of citizens. The ruling party proclaims that it has instituted a democratic system of governance, but has little tolerance for opposition groups or individuals as it incessantly castigates and intimidates other political organizations. The government continues to restrain freedoms of speech, peaceful demonstration, and press. The working class cannot effectively exercise the rights of freedom of association and collective bargaining. Federal as well as regional government security forces act violently against peaceful demonstrators, as the 2002 fatal shootings of several innocent demonstrators in the towns of Awasa, Tepi, Ambo, and Nekemte have proved.[7] Detentions without due process occur frequently.

It is true that the country enjoys more freedom of expression than it did during previous regimes. However, the exercise of such a right is heavily constrained. Independent publications are allowed, although writers and editors are imprisoned or harassed if the state considers their writing subversive. Independent newspapers thus operate under a cloud of persistent threats. The government exercises a virtual monopoly over the airwaves (television and radio) and will not pave the way for the development of competing commercial media that could serve as a watchdog of the government uncovering accountability failures. The ruling party's approach to policy-making is unyielding and shows little or no interest in listening to alternative perspectives or arguments, no matter how constructive they may be. As a consequence, many believe that the ruling party sustains power not because of its adherence to genuine democratic principles, but because it controls the state machinery and the mass media. The state machinery monitors public actions while the mass media invariably perpetuate the state's interest.


## Ineffective Socio-Economic Reforms

For nearly two decades, prior to the early 1990s, the Ethiopian economy suffered from the effects of state-control and the burden of restrictive measures against private entrepreneurship. The new rulers found the country's development lagging by at least 30 years. In 1992, with an extension of a $657 million recovery and reconstruction loan from the World Bank, the new government took a variety of structural adjustment steps to improve the state of the country's economy. This effort included devaluing currency, limiting inflation, reducing the fiscal deficit, reducing restraints on imports, and abolishing government monopolies and price controls (Ottaway, 1999: 73). With these reforms in place, the gross national product grew by an average of 4.7 percent annually from 1990 to 2001, up from 1.1 percent annually during the 1980s. The industrial value added grew at an average of 5.4 percent annually during the same period while it was less than one-half a percent during the 1980s. The agricultural value added also showed a modest growth, a 2.3 percent annual growth over the 1990-2001 periods. This sector, which accounts for 52 percent of the GNP, showed little or no growth throughout much of the 1970s and 1980s (World Bank, 2003: 186). The brightest spot in the economy of the 1990s was the domestic investment, which grew at over 13 percent annually. The export sector also performed quite well, registering a 9 percent or better rate of growth annually over the same period—up from 2.4 percent average growth per year through the 1980s (World Bank: 2001a: 294).

The economy slowed down considerably toward the end of the 1990s largely due to the impact of the senseless border war with neighboring Eritrea between 1998 and 2000. The bloody two and a half year border war diverted precious resources—both human and material—from the acute development needs of the country and reversed the long-term decline of the country's defense expenditures. The war cost Ethiopia nearly three billion dollars, according to a study by the Ethiopian Economic Policy Research Institute. The defense expenditure rose eight-fold, from about $100 million the year before the war to nearly $800 million in 1999/2000.[8] The estimated cost of destroyed public physical and social infrastructure amounted to well over $200 million. Nearly 400,000 people were displaced. Investment—foreign as well as domestic—declined sharply, even though the war was probably only partly to blame. Of the 217 foreign companies' projects that were being approved before the war broke out, only 47 got started by the end of 2000. Of the nearly 14,000 domestic investors who showed interest in investing prior to the war, less than 1,500 launched their projects. The country also lost hundreds of millions of dollars in development assistance from donors who disapproved of the war. Income from tourism declined, as did domestic savings (Bhalla, 2001). Tens of thousands of human lives were lost. All of this taxed the economy very heavily. The gross domestic product annual growth plunged to a mere 0.5 percent in 1997/1998, down from 5.2 percent growth the year before the war erupted.

Four years have passed since the war ended. The economy appeared to be recovering quickly as the gross domestic product grew at 5.2 percent in 1999/2000 and at 8.7 percent in 2000/2001. However, the economy has yet to produce employment opportunities for the country's growing workforce. Nor has the structure of the economy changed. Agriculture remains the predominant means of livelihood, accounting for one-half of the gross domestic product (GDP), 90 percent of exports, and 80 percent of total employment. Agricultural output and productivity have risen, but the sector still lacks adequate production-enhancing inputs, improved tools and physical access to markets, while suffering from unfavorable prices and insecure ownership of land. The sector also suffers from frequent periods of drought and, on average, as many as 4 million people in the 1980s and 6 million in the 1990s needed food assistance every year. In 2000 alone, well over 10 million people—that is, almost 17 percent of the then total population—sought food assistance.[9]

The industrial sector is small but, even though it has been expanding in recent years, its share of contribution to GDP still remains at the 10-12 percent level and employs less than 7 percent of the total labor force. Industry continues to be dominated by four consumer products (food, beverage, textile, and leather) catering almost entirely to domestic markets. Labor productivity and capacity utilization are still quite low. The lack of export markets, shortages in imported inputs and spare parts, and growing consumer shifts to imported goods are major bottlenecks for the sector to perform better than it currently is (EPA, 2000: 1). Import liberalization has also exposed domestic industries to foreign competition. Domestic industries, especially those that produce textiles and leather products, are losing to foreign competitions, and in particular to the Chinese, whose goods, produced at rock bottom wages, have flooded the local markets. Chinese footwear imports have, for instance, resulted in the closure of 20 local shoe factories and the laying-off of 15,000 workers (The African Economist, 2000: 16). Furthermore, while state ownership of the sector remains significant (government owns over a quarter of the major industries that produce two-thirds of the total value of industrial output) (CSA, 2000: 152), it is unlikely that the sector will reach the level of efficiency and productivity that could be achieved under private ownership. In fact, the government approved loans in 2002 in the amount of 250 million Birr to state-owned manufacturing industries facing a serious financial hardship.[10] Taxpayers and private businesses continue to subsidize failing state industries.

By comparison, the contribution of the service or tertiary sector to GDP—now at 37 percent (World Bank, 2003: 190)—has improved measurably on account of increased involvement of the private sector in trading business in the wake of market reforms. The sector grew at an annual rate of 7.7 percent during 1992/1993 to 1997/1998 (World Bank, 2001b: 6), up from 3.1 percent in the 1980s (World Bank, 2003: 186). However, regulatory barriers and lack of access to financial markets, especially for small-and medium-sized business entrepreneurs, continue to stifle the potential growth of this sector. An uneven playing field also constrains the service sector. Critics argue that a handful of private companies that are intimately connected to the political organizations of the ruling party are stifling the competition by acquiring special deals, inside information, unfettered access to resources from public financial institutions, relief from regulatory burdens, and preferential treatment in import and export processing.[11] Such companies are said to have prevented genuine investors from participating in business ventures by having greater access to credit and licenses. Thus, some improvements have occurred in the tertiary sector since 1992, but not for everyone.[12] Overall, the private sector, which can create opportunities for employment and productive growth, experiences many problems including low demand for goods and services, inability to compete with cheap imports, difficulty obtaining land at a reasonable price, bureaucratic incompetence, and unfair competition.

The government maintains that it has put in place (since 1996) the right development strategy to reduce poverty and enhance sustainable and equitable development: Agricultural Development Led Industrialization (ADLI). Through ADLI the government intends to promote agricultural growth in order to end the country's dependence upon foreign food sources to feed its population and to create a market for consumer goods, in turn creating a need for processing industries that generate more jobs. In other words, the strategy envisions the development of small and medium-size agro-industries in close proximity to the rural producers, the promotion of industries that produce agricultural inputs, and the promotion of off-farm employment by supporting rural cottage industries and by developing marketing infrastructure (Gebre Giorgis, 2000: 33). To improve the productivity of smallholder agriculture, the state is to provide extension services, fertilizers, improved seeds, better implements, pesticides, and improved storage facilities to mitigate post harvest losses. The strategy of ADLI thus underscores the forward and backward linkages between industry and agriculture. ADLI assigns important tasks to agriculture, including the attainment of food security, the generation of capital essential for its own growth and industrial development, the supply of human power and raw materials to the industrial sector, and serving as an outlet for industrial products (EPA, 2001: 130). Along with this strategy the government says it has also put into action sector development programs in education, health, road, potable water, and sanitation with the superseding goal of poverty reduction. With agricultural promotion, the government has invested significant funds on road development in rural areas.

The ADLI strategy and the sector development programs sound prudent. However, the experience of the last few years indicates that problems that have been, and continue to be, constraining this development strategy are plentiful. In the agricultural sector, resource limitations continue to hamper the provision of extension services and inputs to a wider farming population. Budget allotted for the agriculture sector remains minuscule; it was only 3.7 percent of total government expenditure in 1999/2000, down from 6.4 percent in 1995/1996 (CSA, 2001: 251). The shortage of foreign exchange limits supplies of most agricultural inputs. When inputs are available, their distribution is affected by the lack of access to motor roads. Demand for inputs is also constrained by high prices, weak extension service, and shortage of credit (Demeke, 1995: 245). The prices of crops have persistently failed to increase at least commensurate to the increase in the price of fertilizer and other inputs. Food security has remained unachieved.[13] The agricultural export sector has not been of any help either. The price of coffee, which accounts anywhere between 65 and 75 percent of the country's export earnings, fell persistently in the international market—down by 70 percent over the last four years alone (Turner, 2002: 4).[14] The export prices of the three main export products of the country fell by 38 percent in 2000 and 35 percent in 2001, generating acute balance-of-payments pressures for the country.[15] All this has hurt too many farmers, and defaults on farm loan payments have become quite common. The land ownership issue remains unclear. Opting to avoid the difficult issues surrounding private and communal land tenures, the government has chosen to maintain the nationalized land system. Farmers have access to a piece of land but do not own it. Thus, farmers cannot use the land as collateral to get a loan to improve their productivity.[16] Farm fragmentation and small-size holdings also contribute to the problem of low farm income. Nearly two-thirds of rural households access only a hectare of land or less. By all accounts, hence, the ADLI development policy has not greatly improved conditions in rural Ethiopia due largely to the fact that the conditions that would enable agriculture to contribute fully to the overall growth process are not yet in place. As long as agrarian poverty persists, poverty-stricken rural peasants will continue to pour into cities and towns—that are already suffering form severe shortages of vital urban amenities—hoping to find a better life. Hence, the government will need to reform the agrarian sector to increase farm productivity. Development in the rural sector, however, cannot happen without complementary improvements in the growing urban areas. Linkage between the industrial and agricultural sectors is minuscule; domestic industries process only 4 percent of the country's agricultural produce.[17]

The post-1991 reforms may have ended economic stagnation and helped growth to resume, but they have not gone far enough to remove all of the impediments that continue to constrain sustained improvement in productivity and outputs in the agricultural, industrial, and tertiary sectors of the economy. As a consequence, opportunities to create more jobs for the ever-growing labor force have been lost. At present, high unemployment and underemployment remain facts of life in both the urban and rural areas. Other critical social problems, such as food security, housing, unemployment, and access to education, health and potable water, still abound. Natural disasters and illness continue to burden millions of people. Poverty remains deep and widespread. Three-fifths of the country's population lives below the poverty line (that is, a dollar a day as estimated by the World Bank).[18] A large proportion of the country suffers from a high level of food insecurity and low levels of food intake. An estimated 40 percent of rural households do not produce sufficient food or income to meet their essential nutritional needs (USAID, 2001: 2). Nearly two-thirds of children under the age of five suffer from chronic malnutrition, one of the highest percentages in the world (Croppenstedt and Muller, 2000: 3). The nation spends no more than 12 Birr a head on healthcare annually. That is only a little more than 7 percent of the total annual government expenditure (Ministry of Health, 2002: 3), a sum that does not meet even the most basic health needs of one-tenth of its population.

Government priorities are still horribly misplaced. Government budget allocation for national defense and public order and security, for instance, grew from 19.5 percent in 1995/1996 to slightly over 50 percent in 1999/2000.[19] During the same period, expenditures for economic services (agriculture, trade, industry, transportation, mines and energy, construction, and tourism) and social services (education and training, public health, labor, culture and sport) together declined from 34.6 percent to 20.2 percent. Budget allocation for all economic and social sectors persistently declined. Public expenditure in the health sector has been awfully low. Not only did this sector receive less than its fair share but also has steadily declined over time, from 5.6 percent of the total government budget in 1995/1996 to 3.5 percent in 1999/2000. In 2000, Ethiopia ranked 189th out of 191 countries in terms of health expenditure per capita (WHO, 2000: 153). Public health service coverage has remained at 50 percent for quite some time. Capital expenditures on economic and social infrastructure were quite substantial, at least initially. Between 1995/1996 and 1997/1998, capital expenditures averaged about 40 percent of the total government outlays, but declined to 32 percent in 1998/1999 and, again, to 20 percent in 1999/2000 (CSA, 2001: 250-251).

The education sector receives a modest proportion of the budget even though its share declined considerably from 16 percent in 1995/1996 to 9 percent in 1999/2000. Although access to education has risen significantly in the past few years, educational indicators show that the challenge remains formidable.[20] Ethiopia enrolls only 57.4 percent of its primary school age children (67 percent for boys and 47 percent girls). Enrollment in secondary school was only 13 percent (14.8 percent for boys and 10.9 percent for girls) (Ministry of Health, 2002: 11). Only a small proportion of those who complete secondary school go on to the university. This is largely due to an overstrained higher education system that can only make room for a small proportion of secondary students.[21] In this situation, there is a heavy bias against female students. Only about 18 percent of all students enrolled in university and professional courses and 21.5 percent of the students enrolled in vocational/technical schools were female in 1998/1999 (Ministry of Education, 1999). Nationwide, repetition and dropout rates are high, especially for female students. Repetition averages 12.5 percent while the dropout rate averages 15.2 percent (USAID, 2001: 7). The huge gender gap in education can be explained by the fact that the Ethiopian culture places greater value in educating boys than girls. Other factors that constrain the education of girls are domestic duties, marriage in early teens, teachers' preferential treatment of boys, and the lack of female teachers to serve as role models—less than 28 percent of teachers are women (USAID, 2001: 7; Sawyer, 1997: 36). Of course, inadequate coverage with educational facilities also compounds the problem.

While the number of schools has increased in recent years, the quality of education has been deteriorating. At primary and most secondary schools books and vital educational resources are almost non-existent. Most school buildings and infrastructure are dilapidated. Classrooms lack adequate chairs and desks and are jammed with 80 or more students, making it almost impossible to focus on individual pupils. Multiple shifts in the same facilities are the norm, owing largely to the shortages of classrooms. About 70 percent of middle school (grade 5-8) teachers are considered under-qualified (USAID, 2001: 7). The federal government's insistence on conducting primary education in various native languages has destabilized the primary school curriculum. Authoring, translating, publishing, and distributing textbooks and other educational materials in 20 or so recognized primary school languages (the Regional State of Southern Nations, Nationalities and Peoples alone uses 11 languages at the primary level) has been, and continues to be, a mammoth challenge to the educational system of the country. With all these self-imposed dilemmas, the country cannot educate the youth effectively (USAID, 2001: 7).

Domestic saving is quite low owing largely to the low income of the population. Ethiopia's per capita GNP is estimated at about $100, much lower than the average $430 per capita income of low-income countries of the world (World Bank, 2003, 234-35). Domestic savings have persistently been falling over the last three decades or so—from an average of 13 percent of the gross domestic product during the first half of the 1970s to an average of 7.2 percent during the 1980s to an average of 6.5 percent in the years 1991/1992 to 1997/1998. Domestic saving covered only about 47 percent of the domestic capital formation in the 1990s (MOWR, 2001: 82), implying that more than half has been supplied by external sources (loans and grants). As a result, the government has accumulated almost $6 billion in debt, which is almost equal to the country's entire annual gross domestic product,[22] up from an average of 32 percent in the 1980s (Ministry of Water Resources, 2001: 82). Balance of payment deficit was $264 million during the same year (World Bank, 2003: 240). In 2001, Ethiopia met all the demands of the IMF and the World Bank and became eligible for debt-relief consideration under the IMF-World Bank program for relieving the debts of the world's heavily indebted poor countries (HIPCs).[23] Under HIPC, Ethiopia is projected to save about $96 million per year until 2021.[24] However, the projected savings will likely not reduce Ethiopia's loans burden to a manageable level. The country continues to devote a significant proportion of its resources to servicing external debt when its social sectors and infrastructure are in dire need of rehabilitation. Between 1998 and 2001 alone, Ethiopia paid a total of $536 million to its foreign creditors,[25] a huge sum of money that could have gone to education, health and the fight against HIV/AIDS. The country pays US $90 million annually in debt servicing alone, almost as much as its total health expenditure.[26]

Ethiopia thus faces a daunting task in terms of poverty reduction. Plummeting coffee prices have increased the burden of servicing the country's debts in the two years since the World Bank and the IMF assessed Ethiopia's financial position (Denny, 2003). The HIPC will not bring permanent relief from debt. Some argue that the debt-relief will not bring relief at all, because Ethiopia will pay as much in debt service payments in 2003-2005 as in 2001.[27] In his detailed examination of the status of Ethiopia's external debt, Ethiopian economist Dr. Befekadu Degefe concludes that the debt burden is so unsustainable that even 100 percent of export proceeds will not be adequate to service the existing debt. Even a total debt write-off for the country will not be able to release enough resources to attain the requisite rate of growth to help achieve the internationally agreed Millennium Development Goals of reducing poverty by half by 2015 (Degefe, 2003: 89-128).[28] At any rate, any reduction in debt payments will be negated as long as the prices of commodities that the country depends upon continue to decline in the world market.

The government has not given sufficient attention to the most critical problems of the country—food security, unemployment, housing, access to education, health, potable water, and sanitation facilities. One would not expect a government that inherited a country after a long, destructive civil war to eliminate poverty immediately. Still, one cannot concede that the current government has dealt adequately with the roots of poverty. First, it has yet to reorient sufficient resources towards poverty reduction. Second, it has yet to put in place a genuine participatory political forum for its citizens: the government does not give civil society organizations a say in the design and implementation of the development agenda of the country.

Ethiopia has failed to attain true economic development not because the country lacks adequate resources, but rather because it lacks the political will to promote the social and economic conditions of the vast majority of the population. Poverty, poor living conditions, and the lack of basic necessities of life continue to impede health and development in the country. Likewise, the country cannot make improvements in the urban physical and human environments without good governance and improvements in the social and economic well-being of the majority of people in the country.

**The Scope and Themes of the Book**

The chapters to follow deal with major urban problems, including inadequate housing, poor access to safe water, healthcare and sanitation facilities, uncollected solid and liquid wastes, and indoor and outdoor pollution. The reason to focus on these problems is that they pose the greatest threat to health in the urban environment. Chapter 2 analyzes the dynamics of urbanization in Ethiopia with special focus on the development of Addis Ababa, the capital city, discussing the factors that have influenced its demographic and spatial growth. This chapter also examines the problems of urban unemployment, housing, transportation, and governance. A major task of this chapter will be to identify and discuss the factors responsible for the urban housing problems and the general urban environmental decay. Chapter 3 deals with the state of water supply and management in urban areas. One of the major tasks of this chapter is to identify and discuss the factors that constrain the provision of adequate safe drinking water to the growing urban population. Chapter 4 discusses the problems of sanitation and waste management in urban areas and their environmental consequences. This chapter argues that inappropriate disposal practices of household, commercial, industrial, and healthcare wastes as well as unsanitary disposal of excreta imperil public health, particularly that of the urban poor. The growth of population, urban agglomeration, transportation, and industrialization all contribute to the steady increase in waste output and pollution. Chapter 5 assesses the extent, sources, and impact of pollution in major urban areas in general and in Addis Ababa in particular. Chapter 6 examines the state of health and health services in Ethiopia. Healthcare services in Ethiopia rank among the lowest in the world. Rudimentary health services reach only about one-half of the population. Much of the rural areas, where more than four-fifths of the country's population resides, have little or no access to modern health services. While most of the health establishments are located in large urban areas, the facilities provide inadequate service because of shortages in adequate health personnel, medical equipment and medicine. One of the principal tasks of this chapter is to show how poor healthcare infrastructure facilities exacerbate the urban environmental health problems. Chapter 7 draws some conclusions towards steps that can be taken to reduce the environmental health burdens in urban areas.

## Notes

[1] According to the World Health Organization, 'environmental health comprises those aspects of human health, including quality of life, that are determined by physical, biological, social, and psychological factors in the environment. It also refers to the theory and practice of assessing, correcting, controlling, and preventing those factors in the environment that can potentially affect adversely the health of present and future generations.' In simpler terms, environmental health 'is concerned with assessing, understanding, and controlling the impacts of people on their environment and the impacts of the environment on them.' Environmental health programs include safe water supply and quality control, domestic and industrial liquid and solid waste management, soil, air and noise pollution control, chemical and food safety, occupational and institutional health safety, housing and settlement safety, vector and rodent control, institutional health, accident and injury prevention and control, and hygiene education. (See World Health Organization (1993), *Global Strategy: Health, Environment and Development: Approaches to Drafting Country-Level Strategies for Human Well-Being Under Agenda 21*, WHO

Document WHO/EHE/93.1 Geneva: WHO; Cited in Annalee Yassi, Tord Kjellstrom, Theo de Kok, and Tee L. Guidotti (1997), *Basic Environmental Health*, Oxford: Oxford University Press, p. 7; Dade W. Muller (1997), *Environmental Health*, Cambridge: Harvard University Press, p. 1; Ministry of Health (1996b), *Study of Water Quality Standard*, Addis Ababa: Hygiene and Environmental Health Department, MOH,; Ministry of Health (1998c), *Hygiene and Environmental Health Policy*, Addis Ababa: Hygiene and Environmental Health Department, Ministry of Health). In this book, the *urban environment* is considered to 'mean the physical environment in urban areas, with its complex mix of natural elements (including air, water, land, climate, flora and fauna) and the built environment constructed or modified for human habitation and activity,' including buildings, transportation networks, water and sanitation facilities, and parks and recreation facilities (Jorge Hardoy, Diana Mitlan, and David Satterthwaite (2001), *Environmental Problems in an Urbanizing World*, London: Earthscan, p. 12).

[2] Poverty defined solely by income threshold is inadequate to understand the full scope of human suffering. Poverty is a multi-dimensional phenomenon. As Ron Shiffman, Director of the Pratt Institute for Community and Economic Development, has argued, poverty exists because of '*low subsistence level*, when there is inadequate income, not enough food or when a family does not have access to decent and affordable housing; *the lack of protection*, when poor sanitation and public health standards deteriorate to the point where disease is rampant, where access to health and medical services is limited... the *absence of affection*, when people are oppressed or fear retribution ..., when authoritarian governments ascend to power people are impoverished, when people are exploited by either an unaccountable public or private sector; *the absence of understanding*, when people are denied access to education...; *the lack of participation*, when people are marginalized and discriminated against' because of their gender, race, beliefs and economic and social status (Ron Shiffman (1997), 'Urban Poverty: The Global Phenomenon of Poverty and Social Marginalization in Our Cities: Facts and Strategies,' Keynote address at the 1995 Salzburg Congress on Urban Planning and Development Conference, New York, p.2).

[3] A 1999-2000 survey indicates that 81.9 percent of the population of the country earned below the international poverty line of a dollar a day (World Bank (2003), *World Development Indicators*, New York: Oxford University Press, p. 58).

[4] Ethiopia is composed of over 80 ethnic/linguistic groups. The major ones are Oromo (35 percent), Amhara (30 percent), Tigray (6-8 percent), and Somali (6 percent). Orthodox Christians constitute between 45-50 percent; Muslims, 40 percent; and Protestants, 5 percent. The remainder adhere to indigenous beliefs (US State Department, 'Background Note: Ethiopia,' Washington, D.C., November 2001. Available at: http://www.state.gov/r/pa/ei/bgn/2859.htm).

[5] There are over 70 political parties in the country. The ruling Ethiopian Peoples Revolutionary Democratic Front (EPRDF) is a coalition of four parties: The Tigray Peoples Liberation Front (TPLF), the Amhara National Democratic Movement (ANDM), the Oromo Peoples Democratic Organization (OPDO), and the Southern Ethiopia Peoples Democratic Organization (SEPDO). Other political parties include: 2 in Tigray, 3 each in

Amhara, Gambela, and Harari, 4 each in Oromiya and Benishangul-Gumuz, 5 each in Afar and Somali, 21 in SNNPR, and 6 national political parties.

[6] For a detailed discussion of grave violations of the elections laws and outright repression against supporters of opposition parties, see Siegfried Pausewang, Kjetil Tronvoll and Lovise Aalen (eds.) (2002), *Ethiopia Since the Derg: A Decade of Democratic Pretension and Performance*, London: Zed Press. In a synopsis of their book, the authors make the following first-hand observation regarding the elections of 2000 and 2001: '...[C]andidates and party workers and supporters of competing parties were harassed and imprisoned already long before the elections. Manipulations with rules—such as delaying the acceptance of candidates' lists of endorsing signatures until after closing dates, or refusing some signature in the last moment and hence disking candidates—were common practice in many places. During the elections, we observed restrictions on party observers, often their exclusion from voting stations; potential voters for the oppositions were prevented from casting their votes; also incidents of ballot boxes being stuffed or stolen after election day were observed... Hard-handed retaliation targeting leaders and candidates of opposition parties followed the elections, particularly in places where a victory of an opposition appeared possible. Arrests, economic sanctions, dismissal from jobs, discrimination in communal services and distribution of relief food were reported from many places.' (*Addis Tribune*, 'Synopsis of the book: Ethiopia Since the Derg: A decade of Democratic Pretension and Performance,' 20 December 2002).

[7] According to the Ethiopian Human Rights Council (ERCHO), in March 2002, at least 128 people were killed after political protests by a local ethnic group in the Tepi region and at least 17 Sidamas were killed when local security forces opened fire on a demonstration in Awasa (United Nations Office for the Coordination of Humanitarian Affairs, 'Ethiopia: Ethnic violence leaves 18 dead in the east,' 13 February 2003, available at: http://www.irinnews.org/report.asp?ReportID=39343). In his October 2002 report to the House of Peoples Representatives, Prime Minister Meles Zenawi admitted that members of his own ruling party, the Ethiopian People's Revolutionary Democratic Front were responsible for the brazen abuse of constitutional and human rights in the Southern Nations Nationalities and Peoples Regional State. This was in apparent reference to the killings of tens of peaceful demonstrators by the police force in Awasa and elsewhere in the region in early 2002 (*Addis Tribune*, 'Meles admits members of his Party in the South responsible for abuse of constitutional rights.' Available at: http://www.addistribune.com/Archives/2002/10/11-10-02/Meles.htm). A May 2003 landmark conference on federalism, conflict and peace building (hosted by the Ministry of Federal Affairs and the German development agency GTZ) also declared that the federal and state judiciary systems in the country have failed to prevent widespread human rights abuses (United Nations Office for the Coordination of Humanitarian Affairs, 'Ethiopia: Judiciary 'failing to stop human rights abuses,' 7 May 2003. Available at: http://irinnews.org/report.asp?ReportID=33930).

[8] After demobilizing troops following the downfall of the Mengistu regime in 1991, the country's defense expenditure had dropped to 3 percent of the Gross Domestic Product and 8 percent of government expenditures. The border conflict with Eritrea shot up the defense expenditure to 30 percent of government expenditure in 1998 (Nita Bhalla, 'War Devastated Ethiopian Economy,' *BBC News Report*, 7 August 2001).

[9] The Federal Democratic Republic of Ethiopia, *Food Security Strategy*, Addis Ababa: The Federal Democratic Republic of Ethiopia, 2002, p. 35. In 2002, the country faced severe food shortages that affected over 14 million people due largely to the failure of the *belg* rains and the late arrival of the main *kiremt* rains. Food shortage is a chronic problem in Ethiopia. Even in a good year 4-5 million people need food assistance. The problem goes beyond the absence of precipitation and includes less than secure land tenure, inability to irrigate land even when water is available nearby, poor infrastructure to haul food from areas of surplus to areas of deficit, and over-dependency on food assistance.

[10] *Addis Tribune*, 'Government approves 250 million Birr for state-owned Manufacturing Industries.' Available at:
http://www.addistribune.com/Archives/2002/11/15-11-02/Government.htm

[11] Critics argue that party-affiliated companies are far from private since they are heavily subsidized by government funds.

[12] In the last decade, wealth has increasingly fallen into the hands of fewer and fewer people and, as a result, the gap between the rich and the poor has widened. In 2000, the richest 10 percent claimed nearly 43.8 percent of the national income, whereas the poorest 10 percent had only 0.7 percent and the richest 20 percent had 60.8 percent of the income while the poorest 20 percent had only 2.4 percent (World Bank (2003), *World Development Indicators*, New York: Oxford University Press, p. 64).

[13] In the 1990s, major food grains grew at an average annual rate of only 0.6 percent while population grew at about 3 percent (EPA (2001), *National Review Report on the Implementation of Agenda 21*, A report prepared for the United Nations Conference on Environment and Development (UNCED), Johannesburg, South Africa, June 2002, Addis Ababa, EPA, p. 63). The average yield for grain crops has improved, but it is still low despite the increasing use of fertilizer and other inputs. As a result the country is still far from being self-sufficient in food and continues to depend on food aid.

[14] Prices of coffee in the world market fluctuate widely, but have recently collapsed due largely to superfluity in the market. Ethiopia is the leading exporter of organic coffee in Africa. In 2002, the country's coffee income dropped by $110 million as compared to the $58 million it was going to save in debt relief during the same year (Jeremy Hobbs, 'Coffee Companies Under Fire as Millions Face Ruin.' Available at: http://www.oxfam.org/eng/pr020918_coffee.htm). More than one million Ethiopian coffee farmers, accounting for nearly 15 million households, have been affected by the continuing fall in the price paid to coffee producers globally (Ellene Mocria, 'Coffee Prices Bitter for Ethiopia,' *The BBC News*, 3 October 2002). Globally, the coffee price has 'fallen 50 percent in the last three years to a 50-year low' (*The Ecologist*, 33, 1 (February 2003), p. 8). There is a wide gap between what coffee producing countries earn and what consumers in the importing countries pay for the same product. At present, the value of retail coffee sales exceeds $70 billion annually, but producers receive only $5.5 billion. (Third World Network, 'African Trade Agenda, Number 3, October 2002.' Available at: http://twnafrica.org). The continuing fall in coffee price has resulted in land use shift in some coffee growing parts of Ethiopia. Many coffee growers in Eastern Hararghe are fast converting their land into *qat* (a shrub whose leaves and buds provide a stimulant with the

qualities of a mild amphetamine when chewed) plantation. Ethiopia exports qat to Somalia, Djibouti, Yemen, and a number of countries in the Middle East (*Addis Tribune*, 'Ethiopian Farmers Replacing Coffee with Qat'). Available at:
http://www.addistribune.com/Archives/2002/10/04-10-02/Ethiopian.htm).

[15] International Monetary Fund/International Development Association, 'The Enhanced HIPC Initiative and the Achievement of Long-Term External Debt Sustainability.' Available at: http://www.worldbank.org/hipc/Long-Term.pdf.

[16] In addition, because farmers work land that they do not own or have ill-defined tenure rights to the land, they hesitate to invest their limited financial resources into environmental improvements because they could lose access in the future. In the Ethiopian context, therefore, secure and stable tenure is imperative to promote food security and environmental protection.

[17] *Addis Tribune*, 'Weak Marketing System Impedes Ethiopia's Competitiveness.'
Available at:   http://www.addistribue.com/Archives /2003/11/21-11-03/Weak.htm

[18] In order to understand the gravity of global poverty, the World Bank uses a poverty line to determine who has a yearly income sufficient to provide for all their basic needs. As Hardoy and others argue, poverty line figures can be easily manipulated to report a preferred finding if there are inaccurate, unrealistic assumptions included in the criteria for the poverty line. Increasing the income cutoff for the poverty line will dramatically increase the number of reported poor people in a region, while a lowering of the income will reduce that number greatly. One of the major assumptions inherent in many poverty line statistics is that the income level needed to meet human needs is uniform within and among countries. This assumption is highly inaccurate. Similarly, urban areas are considerably more expensive to live in than rural areas because they often have inadequate services and higher property rent, and should thus have a higher poverty line. The majority of the poverty line statistic is also based upon the costs of obtaining an 'adequate' diet, a highly subjective criteria. The statistic also fails to account for the income needed to pay for non-food items essential to human needs including adequate shelter, clean water, waste collection, sanitation, transportation, healthcare, and schooling for children. Various studies have noted that 20-30 percent of income may go to rent for housing, while 10-20 percent may go to transportation if housing is on the outskirts of the city. However, these factors alone assume that these basic services are available if families have high enough incomes, an assumption that is largely inaccurate in many urban settings. Political, social, and economic discrimination and unstable income sources are also not considered in the poverty line statistics but play a key role in whether human needs are met (See Jorge E. Hardoy, Diana Mitlin, and David Satterthwaite (2001), *Environmental Problems in an Urbanizing World*, London: Earthscan Publications, pp. 315-317).

[19] For current government expenditure, see CSA (2001), *Statistical Abstract 2000*, Addis Ababa: CSA, p. 251. In 2001, defense expenditures alone amounted to $6.2 billion or 43 percent of total federal government expenditures, up from $2.7 billion in 1992 or 19.3 percent of total federal government expenditures. Armed forces personnel increased from 120,000 in 1992 to 300,000 in 2001. Arms imports in 1999 accounted for 20.5 percent of total imports, the third highest proportion in the world following Eritrea (33.5 percent) and

Saudi Arabia (27.5 percent) (World Bank (2003), *World Development Indicators*, New York: Oxford University Press, pp. 286-88).

[20] According to a recent report by the UNESCO, Ethiopia is listed among 57 less developed countries that will probably fail to meet internationally agreed targets for the provision of education to all children by the year 2015 (BBC News, 'Bleak Outlook for Global Education.' Available at: http://news.bbc.co.uk/2/hi/africa/2464589). On the bright side, adult literacy gained some ground in the 1990s. Among adult males (ages 15 and above), literacy increased from 37 percent in 1990 to 48 percent in 2001. Among young women, literacy increased form 20 percent to 32 percent during the same period. The rate of youth illiteracy also fell in the 1990s. Among youth males (ages 15-24), literacy increased from 52 percent in 1990 to 62 percent in 2001. Among youth female, literacy increased from 34 percent to 50 percent during the same period (World Bank (2003), *World Development Indicators*, New York: Oxford University Press, p.88).

[21] The government's education policy also ends most children's education at 15 or 16, which many in the country find fundamentally wrong and detrimental to the country's human power development.

[22] Oxfam/America, 'Seeds of Hope in a Time of Hunger,' Oxfam, 10 June 2003. Available at: http://www.oxfamamerica.org/emergency/art4636.html.

[23] Ethiopia has a good track record in adopting adjustment and reform programs imposed by the IMF and the World Bank. It has also submitted to the World Bank/IMF a Poverty Reduction Strategy Paper (PRSP) that has been endorsed in 2000.

[24] World Bank, Enhanced HIPC Initiative: Committed Debt Relief and Outlook, Status as of January 2003. Available at:
http://www.worldbank.or/hipc/progress-to-date/relief_and_outlook_Jan03.pdf.

[25] Jubilee Movement International (UK), 'Debt and Famine in Ethiopia'. Available at: http://www.jubileeplus.org/worldnews/africa/ethiopia121102.htm.    Despite    significant increases in aid in the last decade, poverty has continued to increase and the average income of the population has remained unchanged. Because loans are accompanied by interest payments, the government must put the bulk of its economic resources into paying interest on loans instead of into food security programs and healthcare facilities for its impoverished population.

[26] United Nations Office for the Coordination of Humanitarian Affairs, 'Ethiopia: African Ministers Criticize IMF Policies,' 5 June 2003.

[27] Catholic Agency for Overseas Development (CAFOD-UK), *A Joint Submission to the World Bank and IMF Review of HIPC and Debt Sustainability*. Available at: http://www.cafod.org.uk/policy/debtsustainability20020902.shtml;    Charles    Mutasa, 'Canceling Africa's Debt: A Critical Appraisal of the HIPC Process,' pambazuka-news@pambazuka.org, editorial posted on 23 May 2003.

[28] Quotes are from the editors' introductory chapter, pp. 4-5. The Millennium Development Goals have been commonly accepted as a framework for measuring progress in less developed countries of the world. 'The goals focus efforts on achieving significant, measurable improvements in people's lives.' The goals are directed at eradicating extreme poverty and hunger, achieving universal primary education, promoting gender equality and empowering women, reducing child mortality, improving maternal health, combating HIV/AIDS, malaria and other diseases, and ensuring environmental sustainability (Detailed information on Millennium Development Goals available at: http://www.developmentgoals.org).

# Chapter 2

# Urban Growth and Decay

Urbanization is a shift in the distribution of population from rural hinterlands to non-agricultural agglomerated locations. The level of urban development in Ethiopia is quite low. Fewer than one person in five is a city or town dweller. However, the rates at which the country's urban areas are growing are among the highest in Africa. Many social, economic, and environmental problems have accompanied urbanization in Ethiopia and have been ignored for too long. Urban environments in the country have become dilapidated to the extent that they comprise major causes of ill-health. Urban areas are faced with problems of unremitting shortage of housing, inadequate transportation systems and a range of other physical infrastructure, unemployment, and ineffective and corrupt management. Overall, urbanization is taking place with little or no planning controls.

Demographic variables such as population size, vitals, and growth rates have major impacts on the demand of urban social and economic infrastructure. The first section of this chapter reviews the trends and prospects for overall population growth. The second section examines the dynamics and problems of urbanization in Ethiopia, charting rural-to-urban migration and its causes. The last section provides a critical analysis of urban housing problems with a particular emphasis on the impact of public control of urban real estate on housing supply, living conditions, and general urban deterioration. This section argues that improper urban land use policies and inefficient management have caused housing shortages and poor housing and environmental conditions.

## The Demographic Situation

With a population of over 70 million, Ethiopia is the most populous nation in sub-Saharan Africa, second only to Nigeria. The rate of population growth at present is about 3 percent per year. Thus, the country has one of the fastest growing populations in the world. At the current rate, the population of the country will double to about 127 million by the year 2025 (WRI, 1996: 244). The rate of increase is high because the country's birth rate (at 4.5 percent) is among the highest in sub-Saharan Africa while the death rate (at 1.5 percent) has fallen, down from 3.1 in 1950 to 2.3 in 1975 to 1.5 percent in the year 2000. As the death rate continues to decline faster than the birth rate, it is quite likely that the country's rate of natural increase will go up rather than decline (Kidanu, 2000: 8).

Moreover, the country's population will continue to grow for several decades to come because a large proportion of the population has yet to reach childbearing age. Nearly 45 percent of the population is below the age of 15 and 90 percent is below the age of 49 (Kidanu and Rahmato, 2000: 18).

On average, women in Ethiopia now give birth to 6.8 children, slightly down from 7.1 in 1975. The decline in birth rate does not make a dent in the exponential growth of the population. Marriage takes place at a very young age, especially for women. A 1990 National Family and Fertility Survey found that 34 percent all of married women in the country consummated their wedlock at the age of 14 or less; 41, percent between the ages of 15 and 17; and 12 percent between the ages of 18 and 19. Thus, nearly nine in ten women were married in their teens (Kidanu, 2000: 85). In most rural areas and small towns, parents continue to arrange the marriages of their children; a young woman is given away to someone she has never seen or met before. Nearly two-thirds of rural women and about three-fifths of urban women become mothers in their teens (Ezra and Gebre Silassie, 1998: 57).[1] Early marriage is thus a major contributing factor to high fertility and high rate of population growth. By and large, the rapid population growth in the country is 'a symptom that people's survival is endangered because of high infant mortality and lack of security, and because women in particular lack economic and political power within the family and society at large' (Hartmann, 1997: 295).

Family planning programs in the country exist in rudimentary form. They are small and under-funded, catering only to a few residents of major urban places. An estimated 4 percent of the country use contraceptives. According to a nation-wide survey conducted a decade ago, only 4.8 percent of married women and about 5 percent of fecund women not then pregnant used some form of contraceptives. Addis Ababa reported the highest use of contraceptives (17 to 35 percent), higher than in smaller towns (13 to 22 percent) and rural areas (2.1 to 2.3 percent) (CSA, 1990).[2]   The National Population Policy of the Federal Government expects to raise contraceptive use to 44 percent in the year 2015 (Kidanu, 2000: 95). Expatriate and local non-governmental organizations increasingly provide accessible family planning and reproductive health services, especially in urban communities. The Family Guidance Association of Ethiopia (FGAE) has implemented the first community-based reproductive health program in the country. Other NGOs such as Family Health International, the Association for Voluntary Safe Contraception, and the USAID have come together to further strengthen family planning and reproductive services by devising a five-year program that modernizes these services, improves maternal and child healthcare, and provides more resources for STI and HIV/AIDS prevention. These new reproductive health programs have produced a network consisting of 27 clinics and 22 community-based reproductive health programs, which cover 27 urban regions in the country (EPA, 2000: 104). Ethiopia appreciates such efforts; however, family planning alone cannot curb the rapid population growth. Family planning ought to form part of a broad development program that encompasses progress in educating women and improving their status, expansion of healthcare facilities, and provision of clean water and nutrition.

Because of the aforementioned factors, Ethiopia's population is projected to reach 213 million by 2050 (World Resources Institute, 1996: 244) and, possibly, 250-275 million before leveling off, probably towards the end of this century. The country cannot expect to fall to a replacement-level fertility rate (that is, the number of children a couple must have to replace themselves— usually 2.1) before the year 2075, at best. However, the AIDS epidemic may counterbalance the rate of population growth. AIDS has caused a significant increase in the death rate and a significant decline in the average life span in Ethiopia. If this virus remains unabated, life expectancy in the country is projected to fall by at least 12 years, from the current 52 years, by 2015. At present, one in eleven people of 15-49 years of age is HIV positive (MOH, 2001: 9).

## Historical Development of Urban Settlements

Ethiopia has a long history of urban settlement. Early urbanization emerged in the northern part of the country. This is not at all surprising when one considers the great civilization that this region has produced. The earliest evidence of urban settlement, or a nucleated concentration of population, is found in the Aksumite Kingdom (ca. 100-600) that included several towns of varying sizes and ports. At the height of the Kingdom (ca. 500), an estimated 10,000 to 20,000 people may have resided in Aksum, the Kingdom's capital (Butzer, 1981: 488).[3] The city, situated in the north of the country in what is now Tigray Regional State and independent Eritrea, served the Kingdom as a major center of political power, commerce, wealth and religious activities (Phillipson, 1994: 84). The Aksumite Kingdom, one of Africa's oldest civilizations, traded extensively with such distant lands as the eastern Mediterranean, Egypt, South Arabia, and India and exchanged gold and ivory for luxury manufactured goods through its principal Red Sea port of Adulis in the Bay of Zula (Pankhurst and Ingrams, 1988: 9). The Kingdom minted its own coins, developed a writing system (Geez), and erected spectacular palaces, temples, churches, institutions of learning, and stelaes that still stand today (Bekerie, 1997: 19). No one knows how the Kingdom erected these gigantic stelaes of enormous weight: the Kingdom surely had the organizational skills and sophistication to galvanize huge work forces. However, the rise of Islam beginning in the seventh century brought the decline of the Kingdom as it lost its power and primary source of revenue from distant trade. Following the fall of the Aksumite Kingdom, a new ruling dynasty, the Zagwe, emerged sometime during the eleventh century with its power base further southward in Roha—later renamed Lalibala after the most prominent ruler of the line—as its capital. King Lalibala built eleven magnificent monolithic churches, each differently designed and painstakingly carved out of solid rock. The resident population of Lalibala was variously estimated at between 3,000 and 20,000 (Pankhurst, 1965: 68).

From the seventeenth through the nineteenth centuries, towns or nucleated centers of population of varying sizes existed in parts of northern Ethiopia. Gondar, established in the early seventeenth century during the reign of Emperor Fasil (1632-67), was an important political, commercial, and ecclesiastical learning city for over 200 years. Emperor Fasil erected splendid castles, palaces, and bridges, the ruins of which can still be seen in the city today as great symbols of the country's glorious past. Towards the end of the eighteenth century, Gondar was the most populous and important city in the Abyssinian Empire with a population estimated at between 40,000 and 65,000. The city was known to house the largest number of artisans. Its population and significance, however, substantially declined later in the nineteenth century as the central monarchy slowly lost political importance to powerful regional rulers (Pankhurst, 1965: 61; Crummey, 1987: 11).

Other important towns peopled northern Ethiopia during the eighteenth and nineteenth centuries. The most prominent were Adwa, Adigrat, Antalo, Chaliqot, Hawzen, Mekele, Weldiya, Soqota, Mahhdara Maryam, Debre Tabor and Debre Markos. The towns, which all had a permanent population that exceeded 3,000, served as administrative, commercial, and small commodity production centers. They supported diverse specialization and played important exchange roles in the economies of their hinterlands (Crummey, 1987: 14, 20-22). In northeastern Shewa, two towns also came to prominence as seats of government and commercial centers. Debre Berhan (Place of Light) was founded in the fifteenth century by the Emperor Zara Yacob. Though abandoned by its rulers, it became a temporary home for the rulers of Shewa at the end of the nineteenth century. Located east of Debre Berhan, Ankober was the capital of the kings of Shewa until Addis Ababa was founded. Ankober had an estimated population of 15,000 in the early nineteenth century (Ministry of Water Resource, 1997b: 5-13).

Farther east the walled city of Harar was founded as early as the twelfth century on the northern slope of Mount Hakim at the end of the eastern Hararghe Mountains. The city's rise to prominence, however, began during the second decade of the sixteenth century when Imam Ahmad Ibn Ibrahim, also called Ahmad Gragn, ruled it for four decades or so. As the principal capital of the Muslim country in the east, it was a considerable commercial center. Most of its citizens were engaged in trade. The city issued its own currency and was well known for the production of fabulous textiles. A center of Islamic culture, mosques, minarets, and shrines adorned Harar (Pankhurst and Ingrams, 1988: 11). Harari traders and Islamic scholars traveled extensively throughout the Middle East. In the middle of the nineteenth century Harar's population was estimated at between 10,000 and 15,000. But the Egyptians, who captured the city in 1875 but were expelled in 1885, estimated the city's population at between 30,000 and

40,000. Estimates for the period following its forceful incorporation into the Ethiopian Empire in 1887 were not so different (Pankhurst, 1965: 75-76).

Until the reign of King Menelik II (1889-1913), the growth and expansion of Ethiopian urban places was essentially low. This was due primarily to the continued military and political confrontations among regional kings and chieftains and the absence of a consolidated political power structure necessary for the formation and perpetuation of urban agglomeration (Wolde Mariam, 1972: 185). By the end of the 19$^{th}$ century, Menelik had established a territorially defined empire state with Addis Ababa, founded in 1886, as the permanent capital. During Menelik's rule a number of urban settlements emerged and existing agglomerations strengthened, especially in the southern half of his empire. After his warlords annexed the region, they established garrison towns (*ketema*) to control and administer the conquered population and to exploit their resources. Most of these garrison towns and those that had existed prior to the occupation eventually developed into permanent administrative centers and commercial and regional market centers (Horvath, 1969: 77-88). With increased specialized functions and the central government's bureaucratic presence, their population steadily grew.

At the beginning of the twentieth century many urban centers developed in response to the creation of a modern transport network. A major development was the construction of the Addis Ababa-Djibouti, a railway 781 kilometers long. The construction of the railway, which began in 1897, stimulated the emergence and growth of several urban centers. By the time the railway reached Addis Ababa in 1917, over thirty stations had been established along the line. Many of these stations (including Akaki, Kaliti, Dukem, Mojo, Bishoftu, Adama, Awash, Dire Dawa) later grew into urban centers (Wolde Michael, 1967: 125). For instance, founded in 1902 to serve the rail link from Djibouti to Addis Ababa, Dire Dawa is today the second most populous city in the country.

The Italian 1936-1941 occupation also directly spurred the expansion of urban settlements elsewhere. The Italian colonialists needed to develop an infrastructure for the systematic exploitation and control of the country. During the five-year occupation, the Italians initiated extensive all-weather road building programs in the country, allocating over four-fifths of their East African budget to road building (Baker, 1994: 154). They built several roads connecting the capital with a number of regional administrative centers. In the process, many towns, and later cities, grew up alongside the roads. The Italians not only left their imprints on the urban landscape by establishing new towns, but also stirred the growth of urban centers that had been in existence before their occupation of the country. In the latter case, they added new functions (such as army garrisons, small factories, and grain mills) to expand the economic base of the urban centers (Horvath, 1968: 42-48).

During the decades following the restoration of political independence, the pace of urbanization was slow, although steady. Ethiopian urbanization took three forms. First, the number of small urban centers (mainly administrative and commercial and trade centers) increased. Second, a rural exodus caused the urban population to increase. Third, there was a concentration of urban growth in the larger towns with Addis Ababa emerging as the dominant city.

Industry has not played a role in the origin of Ethiopian urban centers, nor has it served as an economic base for their growth. Instead, urban growth has been largely dependent on commercial activities and administrative functions. However, burgeoning industrialization has played a part in stimulating the growth of some cities such as Addis Ababa, Dire Dawa, Bahr Dar, Akaki, Adama and a few others.

## The Development of Addis Ababa

Unlike most African capital cities, Addis Ababa is centrally located. The third highest capital city in the world, it lies in the central highlands at an altitude of 2,400 meters, making it a comfortable climate. The city spreads over five hills that slope down from north to south, cut by a number of steep sided valleys of rivers and streams—the major ones being the Qabbana, Qachane, Quorum, and Little Akaki. For several centuries the Oromo cultivators and cattle keepers inhabited it almost exclusively, calling it Finfine. When Menelik abandoned his temporary headquarters in northern Shewa and moved to Finfine, the riches of Finfine and its surroundings enthralled him: immense agricultural land, water, and forest resources. It is said that nearly half a century before him, his grandfather, King Sahle Silassie, was also enchanted with the environs of Finfine and prophesied: 'Oh, you land! A day will come when my grandson will make a town out of you' (Foucher, 1987: 233). Menelik first established his military headquarters at the Entoto Hills, some 2,800 meters high. However, after a brief period of encampment the Entoto Hills proved to be poorly endowed with water and wood supplies and 'very cold and windy' (Pankhurst, 1962: 35). Menelik—attracted by the mildness of the climate, the availability of more firewood, and the presence of thermal springs—moved southward to the gentler, lower slopes about eight or so kilometers from the steep upper slopes of the Entoto Hills.[4] In 1886, he erected a spacious palace on a hill adjacent to the thermal springs, signaling the birth of Addis Ababa (the name means 'new flower') as a permanent capital.

Less than a decade after it was founded, the new capital nearly fell apart. Faced with the depletion of wood for fuel and construction, Menelik temporarily moved his capital to Metcha (later renamed Addis Alem by Empress Taitu), some thirty kilometers to the west of Addis Ababa. But the discovery of coal not far from the capital and, perhaps more importantly, the importation of the fast-growing eucalyptus tree (*Eucalyptus globulus*) from Australia in 1905 saved the new capital from being abandoned and rendered insignificant, like the capitals before it (Pankhurst, 1968: 246; 1962: 51). Menelik's proclamation ordering the replanting of imported trees was scrupulously heeded and the inhabitants of the city 'grew eucalyptus trees all over the capital' (Horvath, 1968: 313). Foreign residents of the city, embassies, or legations participated in extensive eucalyptus planting as well (Pankhurst, 1962: 51). Within a short period, the planting of the eucalyptus trees spread widely to areas far beyond the immediate vicinity of the capital (Kebbede, 1992: 85; Horvath, 1968: 13-19).[5]

The early patterns of settlements in the city were rather scattered. In the first decade of its development, the city was just a cluster of villages or *safars* in which prominent war chiefs and their families and followers settled on land granted by Menelik. Class did not segregate *safars*. The nobility and their soldiers, kinsmen, and servants lived next to each other in a sort of hierarchical structure. Each cluster of settlement grew up around spacious residences of the nobility and gradually expanded into empty tracts of land thereby shrinking the distance between the *safars*. The *safars* initially housed a population that was predominantly soldiers and servants or slaves of the nobility. In the late 1890s, especially after the victory against the Italians at Adwa, the city rapidly acquired a considerable number of civilian populations with increased commercial activities and a growing presence of foreign communities. Menelik's 1907 decree of permanent occupation of Addis Ababa sealed the city's status as the nation's capital (Assen, 1987: 80).[6]

Addis Ababa experienced profound changes during the first three decades of the twentieth century. The construction of housing burgeoned, partly fueled by the abundance of eucalyptus trees and the importation of corrugated iron roofing. With the establishment of a municipal administration, public works picked up quickly. Several public buildings (hospitals, schools, churches and government institutions), featuring European style architectural designs and forms, were erected (Habte Mariam, 1987: 201-202). Italian prisoners captured at the Battle of Adwa in 1896 were used to build roads and bridges that linked the *safars* of the nobility with the palace and with each other (Zewde, 1987: 46). Notable establishments like the Taitu Hotel (1906) and the Bank of Abyssinia (1908) came into the urban scene. Foreign residents (Italians, Armenians, Greeks, and Indians) introduced a number of small-scale industries, such as brick factories, furniture shops, and bakeries. A pipe water system that primarily served the palace and some of the nobility was introduced. Hotels, restaurants, taverns, retail shops, flourmills, and butcheries sprang up in large numbers. Migration into the city from the surrounding countryside and far away places of the empire increased.

The completion of the Addis Ababa-Djibouti railway line in 1917 and the building of a railway station were major catalysts to the city's southward expansion. The station provided employment to a considerable population of wage laborers and stimulated the development of nearby residential dwellings (Chapple, 1987: 153). The railway line became a major point of entry into the city for imported products (including cotton cloth, foot wares, household goods, and kerosene) and, consequently, made the Arada Market, located on a slope only two kilometers from the railway station, a thriving commercial center. The railway line may have also hastened the arrival of motorcars in the 1920s that gradually transformed the political character of the city into an expanding commercial center (Ahderom, 1987: 249). On the eve of the Italian occupation in 1935, the city's population had reached 100,000 (Chapple, 1987: 143), with the built-up area covering about 1,863 hectares (AAMPRO, 1999a: 17).

Early in their occupation of the city, the Italians undertook a few public works, for they were uncertain about retaining the city as the capital. The Italian occupiers (frustrated by the hit and run activities of Ethiopian patriots from hid-outs in the forests and ravines of the city) had considered moving the capital elsewhere, but later abandoned this idea because Addis Ababa provided prestige and a great symbol of conquest (Pankhurst, 1987: 119-122). The latter years of the occupation saw the construction of roads and bridges and the erection of several government buildings, cathedrals, shops, hotels, cinemas, small industries, residences, and a few schools and hospitals. In 1937, the Italians introduced the first master plan of the city that was based on the principle of racial and economic segregation. Nearly all the native citizens were evicted from their houses and expelled from their land, forced to move to a designated area west of the city's center (later to be known as Addis Ketema) with its own central market place or the *Merkato*. Where European residential quarters were created, native dwellings were burned to the ground, except for the larger ones (Pankhurst, 1987: 131; Habte Mariam, 1987: 204). The plan shifted the main commercial and political centers southward to the level grounds and created an industrial zone in the southwest near the commercial center, the railway station, and residential areas designated for the native citizens (Zewde, 1987: 50; Habte Mariam, 1987: 204).

By the time the Italian occupation ended, the city covered about 70 square kilometers and its population had reached 150,000 (AAWSA, 1984: 4). After independence the direction of the growth of Addis Ababa continued southward into areas of lower altitude: the hills running across the north of the city restricted the possibilities of northward expansion. The building of airports (first at Lideta and later at Bole), the emergence of various international organizations, and the growing number of embassies and chancelleries played important roles in advancing the city's spatial growth toward the south and southeasterly directions (Zewde, 1987: 45).

In the post-Italian period the government undertook a number of steps in an attempt to reorganize the internal structure and to guide the growth and development of the city. First, *weredas* (districts) replaced *safars* as administrative units. Ten *weredas* were created: Arada, Bole, Entoto, Gafarssa, Gulale, Maychaw, Mekakelegna, Qaranyo, Teklehaimanot, and Yekka. Second, in 1954 Addis Ababa was granted a legal charter as a municipality, entitling it to some measure of self-rule or autonomy (Desta, 1996: 29). Third, a master plan was initiated in order to induce planned growth and thus end the spontaneous and haphazard growth that had characterized the city ever since its inception. Sir Patrick Abercrombie, the renowned planner of Greater London, developed the first post-independence master plan for the city in 1956. His plan would have divided the city into different functional zones: residential, commercial, industrial, recreational, and governmental zones (Pankhurst, 1957: 37). In addition, Abercrombie envisioned street networks that radiated out from the city center and merged into an encircling ring road. In 1959, another British team of planners, Bolton Hennessy and partners, were commissioned to improve on Abercrombie's plan to emphasize future growth of the city in terms of physical size and population. Their plan proposed a metropolitan region with five satellite towns surrounding Addis Ababa (AAMPRO, 1999b: 2-11). However, neither Abercrombie's nor Hennessy's plans, which would have cost millions of dollars to produce, materialized for a number of reasons. First, the master plans lacked legal enforcement; second, the city lacked adequate resources and management capacity; and third, perhaps most importantly, the entrenched urban landed aristocrats resisted the plans for fear of conceding some of their vast private holdings.

The population of the city grew rapidly in the post-war period as a result of natural increase and substantial migration. The principal sources of employment—civil administration and commerce—grew rapidly. During the 1950s and 1960s, expansion of the industrial and tertiary sectors came about on the already existing locational advantages of the city. These sectors provided additional employment opportunities. As the nation's capital, Addis Ababa increasingly enjoyed a disproportionate amount of the country's economic and social infrastructure (schools, hospitals, financial and banking institutions, transportation and communication networks, and manufacturing industries); the capital therefore attracted migrants.

The population of the city increased by about two-thirds between 1970 (795,500) and 1984 (1.4 million).[7] Its share of the urban population, however, declined slightly from 31 percent in 1975 to 30 percent in 1984. This was probably due to the fact that secondary urban centers became more important administrative centers during the Derg regime. Mobility restrictions imposed by the regime in an effort to curtail inter-urban as well as rural to urban migration may have also contributed to the relatively slower growth of the city's population during this period (Kebbede, 1992: 41-43). The city's population nearly doubled to about 2.6 million between 1984 and 2000, but its share of the urban population declined to 27 percent, down from 30 percent in 1980 (World Bank, 2003: 156).

During the decades of 1980s and 1990s, the city continued its fragmented and haphazard physical expansion by annexing peripheral settlements. Major areas of expansion during the two decades included Bole-Megenagna, Old Airport-Mekanisa, Gotera-Nefas Silk, Gulele-Kolfe, Kotebe, Gerji, Qaranyo, Kaliti, and areas along the Jimma road (AAMPRO, 1999a: 17-18). Through this expansion, the size of the city increased to one and one-half times. No planning framework of any sort guided this development, even though the city had produced its most recent Master Plan in the mid-1980s. This Master Plan was to serve the city for a 20-year period ending in 2006, but the government that commissioned it was overthrown in 1991 before any of the proposals in the plan were put to use. In 1994, the Addis Ababa City Administration formally approved the Master Plan, but it took another eight years to be enacted into law. In the mean time, the city continued its typical random spatial growth. Hence, facts on the ground have changed quite immensely since the appearance of the Master Plan in 1986, and, as a result, most of the plan's intended goals were irrelevant to the situation in the year 2000. The Addis Ababa Master Plan Revision Project Office is now revising the Master Plan for possible future use (AAMPRO, 1999b: 2-11). Neither this nor other past master plans ever solicited the participation of urban communities during the planning processes.

For the most part, the city is not residentially segregated by income. In most neighborhoods the rich, the middle class, and the poor often live side by side. The disparity in housing quality between these groups is conspicuous. The residences of the rich are often built with bricks and cement and are quite spacious with several rooms. Service quarters, with two to three separate rooms, line up behind the main residence and stone or concrete walls and a steel door enclose the whole premise. A security guard always sits by the gate safeguarding the properties of the more affluent households.

The 1986 Master Plan had defined the core city areas covering about 202 square kilometers and areas beyond the core were to be added as required. However, in 1995, following the introduction of the regional government, the city's size was more than doubled to some 540 square kilometers, incorporating the satellite towns of Kotebe and Akaki and other rural settlements. The developed areas comprise about 289 square kilometers (54 percent) of the city's size. Residential areas occupy 53 percent of the constructed area while transportation accounts for 28 percent and industrial land use amounts to about 14 percent. The remaining 251 square kilometers of land constitute an expansion area (Tadesse, Bantayehu, and Gebre Mariam, 2000: 28; AAMPRO, 2000b: 35). Even with all this expansion the city lacks parks and recreational facilities. Parks and open spaces figured prominently in the city's revised Master Plan. However, the Plan has been overtaken by the overwhelming reality of expeditious growth even before it became a reality.

## Urban Population Growth and Distribution

Extensive urbanization in the industrialized countries involved a process that took well over a century. The protracted process led to enormous social and economic transformation. Heavy industrialization induced massive inflows of people from rural hinterlands. The condition of urbanization in the developing nations of the world does, however, differ from that of the developed countries. First, urbanization in the developing countries outpaces and exceeds the level of growth that occurred in the industrialized countries during their earlier periods of urban-industrial expansion. Second, the rate of urban growth in the developing countries often outpaces the rate of growth of industrialization. Third, urban growth in the developing countries is induced not only by immigration from the rural hinterlands, but also by continued high rates of natural increase and concomitant reductions in the mortality rate due to public health services and vaccination efforts (Palen, 1981: 335).

Various factors lead to rural-to-urban migration. For the most part, rural migrants are attracted to urban areas by the expectation of employment and better earning opportunities as well as the pervasiveness of social, economic, political, and environmental constraints in the rural areas and small villages. Factors such as poverty, landlessness, diminishing agricultural income, drought, famine, and war constitute major problems that force rural people to migrate. The latter three factors have been unrelenting tribulations for millions of rural inhabitants in Ethiopia over the last four decades.

Ethiopia is currently one of the least urbanized countries in the world, even though, as noted earlier, the first urban settlements date back at least 1,500 years. In the early 1940s, less than 3 percent of its total population lived in urban areas defined by the CSA of the federal government (Wubneh and Abate, 1988: 138). This proportion increased to 8.5 percent in the mid-1960s, to 11.4 percent in 1984 (CSA, 1984), and to 15 percent in 2000 (CSA, 2001b). The average annual rate of growth of the urban population was 9.6 percent between 1967 and 1984; though it declined to 6.7 percent between 1984 and 2001, it was still one of the highest rates of increase in urban populations in Africa. At this rate of growth, the proportion of the urban population is expected to rise to 30 percent in 2025, up from just over 9 million in 2001 to 38 million in 2025.

Urban growth rates vary between large and small urban areas (Table 2.1). Urban areas with a population of 50,000 or more grew at an average rate of 7.6 percent per year between 1967 and 1984 and at 5.5 percent between 1984 and 2001. During the same period, on the other hand, urban areas with populations less than 50,000 grew at 12.8 percent and 8.1 percent, respectively. Growth rates also varied between 3.5 and 12.7 percent among the largest urban areas. Jijiga, Arba Minch, and Awasa had the highest growth rates (in excess of 10 percent annually) over the 1984-2001 period while Bahir Dar, Adama, Nekemte, Shashamene, Dire Dawa, Mekele, Jimma, and Asela grew between 6 and 9 percent annually. Gondar, Dessie, Bishoftu, Addis Ababa, Debre Markos, and Harar registered the smallest growth rates (3.5 to 5.5 percent annually) during the same period. Bahir Dar, Awasa, Jijiga and Arba Minch registered growth rates of 20 percent or more per year during the 1967 and 1984 period.

In general, urbanization in the country is characterized by the prevalence of a high proportion of small towns and the dominance of a few larger urban areas, especially the nation's capital. There were 641 urban areas with populations of 2,000 or more in 2001. Of these, 333 (52 percent) had populations of 5,000 or less, 243 towns (37.9 percent) had between 5,000 and 20,000, 43 (6.7 percent) had between 20,000 and 50,000, and only 22 (3.4 percent) had populations exceeding 50,000. There were only ten urban areas with a population of 100,000 or more, but they had 41.4 percent of the total urban population in 2001.

**Table 2.1 Population Growth of Urban Centers with Population 50,000 and Above, 1967 – 2001**

| | Pop. 1967 | % Urban Pop. | Pop. 1984 | % Urban Pop. | Pop. 2001 | % Urban Pop. | Growth Rate 1967-1984 | Growth Rate 1984-2000 |
|---|---|---|---|---|---|---|---|---|
| Addis Ababa | 644,190 | 38.8 | 1,423,182 | 32.4 | 2,570,000 | 27.5 | 7.1 | 4.7 |
| Dire Dawa | 50,733 | 3.1 | 99,980 | 2.3 | 227,494 | 2.4 | 5.7 | 7.5 |
| Adama | 27,812 | 1.7 | 77,256 | 1.8 | 180,537 | 1.9 | 10.4 | 7.9 |
| Gondar | 30,734 | 1.9 | 80,675 | 1.8 | 156,087 | 1.7 | 9.5 | 5.5 |
| Dessie | 40,619 | 2.5 | 71,565 | 1.6 | 135,529 | 2.5 | 4.5 | 5.3 |
| Mekele | 23,105 | 1.4 | 62,668 | 1.4 | 134,996 | 1.4 | 10.1 | 6.9 |
| Bahir Dar | 12,463 | 0.8 | 54,723 | 1.3 | 134,062 | 1.4 | 19.9 | 8.5 |
| Jimma | 30,580 | 1.8 | 60,218 | 1.4 | 125,569 | 1.3 | 5.7 | 6.4 |
| Bishoftu | 22,055 | 1.3 | 55,657 | 1.3 | 103,569 | 1.1 | 8.9 | 5.1 |
| Harar | 42,771 | 2.6 | 63,070 | 1.4 | 101,000 | 1.1 | 2.8 | 3.5 |
| Awasa | 5,573 | 0.3 | 36,367 | 0.8 | 98,917 | 1.0 | 32.5 | 10.1 |
| Jijiga | 4,030 | 0.2 | 24,716 | 0.6 | 78,167 | 0.8 | 30.2 | 12.7 |
| Sahasha-mene | 7,837 | 0.5 | 31,884 | 0.7 | 73,563 | 0.8 | 18.0 | 7.7 |
| Debre Markos | 21,530 | 1.3 | 41,138 | 0.9 | 68,598 | 0.7 | 5.4 | 3.9 |
| Asela | 13,886 | 0.8 | 32,954 | 0.8 | 66,836 | 0.7 | 8.1 | 6.0 |
| Nekemte | 12,691 | 0.8 | 28,703 | 0.7 | 66,732 | 0.7 | 7.4 | 7.8 |
| Arba Minch | 2,890 | 0.2 | 20,703 | 0.5 | 55,122 | 0.6 | 36.3 | 10.7 |
| Debre Berhan | 9,188 | 0.6 | 25,637 | 0.6 | 53,889 | 0.6 | 10.5 | 6.4 |
| Soddo | 10,842 | 0.7 | 24,278 | 0.6 | 51,888 | 0.6 | 7.3 | 6.7 |
| Total | 1,013,529 | 61.1 | 2,315,374 | 53.0 | 4,484,667 | 48.0 | 7.6 | 5.5 |
| Other Towns | 646,581 | 38.9 | 2,050,703 | 47.0 | 4,859,091 | 52.0 | 12.8 | 8.1 |
| Grand Totals | 1,660,110 | 100.0 | 4,366,077 | 100.0 | 9,343,758 | 100.0 | 9.6 | 6.7 |

Source: Central Statistical Office (CSO) (1967), *Statistical Bulletin*, No. 1, Addis Ababa: CSO; CSO (1984), *The 1984 Population and Housing Census*, Addis Ababa: CSO; CSA (2001), *Statistical Abstract, 2000*, Addis Ababa: CSA.

Two factors account for the rapid growth of Ethiopia's urban population—natural growth and rural-to-urban migration. Of these two factors, the latter plays the more significant role. Large urban areas in Ethiopia owe much of their growth to immigration. Generally, natural population growth accounts for two-fifths of the population increase of large urban areas; net migration from rural areas and small towns accounts for three-fifths of the increase.

Both economic and non-economic factors explain the high rate of rural-to-urban and small town to large town migration. The economic factors are, however, predominant and more pertinent. The growth of the urban economy in the country has largely evolved around a few urban centers. Major industrial and commercial enterprises are concentrated in a few urban centers. Major infrastructure facilities—water supply, electricity, telephone systems and other social services such as education and health—are concentrated in a few urban areas. Thus there exists a marked pattern of inequality with a few urban centers developing out of proportion to the vast rural areas and small towns in the country. For decades the economy of the rural sector failed to expand fast enough to support the ever increasing population and consequently people were pushed more and more to urban areas.

The large migration of people from the less developed areas of the country to the more developed urban centers reflects the 'push and pull' forces operating as a result of unbalanced development in the country. Rural migrants are pushed from their lands because of the diminishing returns on their overused land and the absence of alternative non-farm employment nearby. The belief that urban areas offer considerable employment opportunities and social amenities acts as a 'pull' factor. This 'pull' of the city supplemented by the 'push' from the countryside have so far accounted for the greater part of the country's urban population growth, especially in the larger urban centers. Larger urban areas will continue to grow rapidly because they can promote far more social and economic infrastructure development than smaller towns and rural places. Even though conditions in urban areas are poor, people from rural areas will keep migrating. This situation largely reflects the appalling lack of social and economic development in the countryside.

In nearly every urban area in the country, both large and small, women outnumber men. The magnitude of rural-to-urban migration also favors women, for a number of possible reasons. In rural areas the distribution of the means of livelihood (land, for example) tends to favor men over women. Employment opportunities for women in rural areas are also in short supply. On the other hand, opportunities to make a subsistence living are much better for them in urban areas; for example, employment in domestic work, brewing local beer, petty commodity trading, and prostitution (Baker, 1994: 160-61). Abusive relationships, social pressures to marry at a very young age or unwillingness to tie into an arranged marriage may also cause young women to run away to urban areas. In general, many young rural women seek escape from the drudgery of agricultural work in rural areas and wish for the higher wages and greater access to government services that urban areas are assumed to provide.

By all measures Addis Ababa is overwhelmingly large in comparison to the country's other urban areas. It has a population 11 times the size of that of the second largest city, Dire Dawa, and more than twice the combined population size of the next 9 largest cities. The capital dominates the other urban areas economically, politically, and socially. The initial advantage enjoyed by the city continues to make it more attractive for the concentration of commercial and industrial activities. A heavy concentration of commerce and industries in the capital means the concentration of accompanying services: transport, water, electricity, health, financial, and educational services. Such services, in turn, attract people—skilled and unskilled—from all over the country. The 1994 census indicated that about 42 percent of the city's residents were born outside the city. Almost two-thirds of these immigrants were from rural areas and the remaining came from other urban areas—large and small. The term 'primate city' is applicable to describe the pattern of urbanization in Ethiopia. The capital city is by far the leading economic power, the seat of the federal government, the cultural, financial and manufacturing center, and the nucleus of transportation and communication networks. With less than 4 percent of the total population of the country and 27 percent of the urban population (World Bank, 2003: 156), Addis Ababa accounts for over 31 percent of the workers in the distribution and service sectors, about 20 percent of the total wholesale and retail establishments (Ayenew, 1999: 25), 47 percent of the country's manufacturing employments, 68 percent of the manufacturing establishments (Gebre Egziabher, 1997: 100), 21.8 percent of the doctors (MOH, 2001: 25), 37.2 percent of the hospital beds, and 77 percent of the registered vehicles (Tadesse, et. al, 2000: 40). In the fiscal year of 2002/2003, Addis Ababa received 72.3 percent of all investment projects in the country (Ethiopian Economic Association/Ethiopian Economic Policy Research Institute, 2003: Table 6-3).

Ethiopia has had a policy of decentralization since the early 1990s with the establishment of a federal arrangement for the nation's major linguistic groups. Nine regional urban centers and 53 zonal urban areas have become the centers of social, cultural, and political activities (Gebre Egziabher, 199: 701) for their respective political jurisdictions. As a result, many of these urban centers have assumed vastly increased bureaucratic functions and could serve as future growth centers for their surrounding hinterlands with improved service and transport facilities. The city of Harar is the seat of the autonomous Harari Regional State. Dire Dawa, the second largest city in the country, has yet to be granted self-rule. Both the Oromiya and Somali states lay claim to this predominantly commercial and multi-ethnic city. With 48 percent of the city's population, the Oromo ethnic group is by far the majority, followed by the Amhara (28 percent) and the Somali (14 percent) (CSA, 1999: 44). Nevertheless, the politics of the controversy has left the city under the federal government control and munificence without any form of legal status.

The decentralization policy has transferred the burden of creating jobs and building social amenities to state governments. The federal government policy encourages educated and non-educated people to seek employment in their own regional states. At present, the job opportunities available at the state level exist mostly in the state bureaucracies; the chances of finding employment in the private sector are minuscule. Obviously, by virtue of their large size and multiple functions, regional capitals tend to attract employment seekers from their own hinterlands and will, hence, grow much faster than any other urban areas within their respective region. If that has to occur, a structure of intermediate-size cities might emerge in the country. However, such intermediate-size cities may become dominant cities within their respective regional states. For instance, Awasa, Bahr Dar, and Mekele have already become the most dominant and fastest growing cities within the regional states of the SNNPR, Amhara, and Tigray, respectively. As administrative centers with relatively good physical infrastructure, these cities are overwhelmingly favored for investment and the development of cultural services. One finds, hence, government, education, and commerce disproportionately located in these principal cities.

## Major Urban Problems in Ethiopia

The preceding discussion gives the backdrop against which one can appreciate the nature and magnitude of urbanization problems in the country. Incontrovertibly, each urban area has a distinctive set of problems that necessitate more specific and thorough analysis than what the discussion below promises to provide. However, one can assume that most urban areas in the country share the same types of, or identical problems, but to varying degrees, and examples from some major urban areas in essence exemplify the conditions in other urban areas.

### The Urban Unemployment Problems

Urbanization that occurs without adequate industrialization cannot produce sufficient employment or secure wages. Unemployment and underemployment pose major problems to urban areas because not enough formal sector jobs exist to absorb the growing productive forces in cities and towns. In almost all urban areas the rates of increase in employment opportunities do not keep pace with the rates of increase in people seeking employment. This dilemma is particularly exacerbated by the continued influx of rural-to-urban migrants most of whom are young people, who, having reached a certain educational level, aspire to take up jobs in towns and cities. These growing segments of young people often look for wage-earning jobs in the modern sector and, in many cases, government jobs. Unfortunately, the number of job seekers far exceeds the absorptive capacity of these forms of employment. Most young job pursuers have no skills to sell, and most urban areas lack the economic base to support such unemployed or unemployable people. Many young people remain idle for years while looking for a job that matches their aspirations. In the mean time they live as freeloaders on the skimpy incomes of their relatives. Worse yet, some are involved in criminal activities to survive.[8]

The early 1990s were especially tough for employment seekers in urban areas. The transition from a state-mandated economy to a partially free economy was harsh for people who were on the government payroll. The full implementation of the IMF-World Bank structural adjustment program resulted in the elimination of tens of thousands of superfluous public sector jobs. Most of the disbanded military and security personnel from the previous regime ended up in the streets of Addis Ababa and other major urban areas. The school systems turned out several thousands of skilled and unskilled workforces each year that roamed the streets looking for jobs that did not exist. Thousands of refugees returned home. The rulers of the then de facto independent Eritrean state deported tens of thousands of Ethiopians, including a large number of Eritreans whose only crime was working as civil servants for the previous Ethiopian regime or being spouses of resident Ethiopians.

Even though reliable quantitative data are hard to come by for each urban area, there is some evidence that a significant proportion of the urban populations have no access to adequate means of livelihood. A 1994 socio-economic household sample survey of seven major cities with a population of 100,000 or more showed a high level of unemployment. For instance, Addis Ababa had 41.7 percent of its workforce unemployed; Awasa, 36.1 percent; Bahir Dar, 31 percent; Dessie, 38.5 percent; Dire Dawa, 45.1 percent; Jimma, 34.3 percent; and Mekele, 29.6 percent (Department of Economics/Addis Ababa University, 1995: 16). In all these cities, more women were unemployed than men. The majority of the unemployed were young: about 68 percent were between the ages of 16 and 25 and 90 percent were below the age of 35. Over 80 percent of the unemployed were first time job seekers. Nearly 55 percent had completed high school while about 23 percent had dropped out.[9] About 10 percent had some elementary education and 7 percent had a post-secondary education. Those with no literacy at all were less than 4 percent. Nearly one-half of the unemployed were interested in public jobs (at a time when public jobs were drastically cut) while close to one-quarter hoped for private sector employment. Less than one-third reported not to mind taking any job offered to them (Department of Economics/AAU, 1995: 85-87).

Job placements in the urban formal sector greatly favor men over women. Of the 119,215 urban registered job seekers between 1995 and 1999 (67,101 men and 52,114 women), only 11 percent got placement. Women accounted for only 20 percent of these job placements. Only 5 percent of women job seekers got placements whereas 16 percent of men got placements (EPA, 2001: 20).

**The Informal Sector**

The overwhelming proportion of the urban labor force in Ethiopia, especially the poor, works in informal sector employment—the production and exchange of petty commodity goods and services outside the regulated formal market sector.[10] In the last decade or so, informal activities have grown rapidly in urban areas partly because of increasing urban poverty and the lack of employment opportunities in the formal sector. Informal sector jobs include a very wide range of small-scale activities such as selling cooked and uncooked food, apparels, wood for fuel, and providing services such as hauling, domestics, and prostitution. Most are individual or family operated activities that are not legally registered or licensed. Informal sector workers tend to have little or no access to institutional credit, public service, and organized market space. The government does not recognize or regulate them. Some operate from a fixed location, but many do not. A 1997 CSA survey in 48 major urban areas found about three-quarters of a million people were engaged in informal sector employment of which almost two-thirds involved women. Four-fifths of the activities were owner operated. Forty-seven percent of these activities involved manufacturing of petty commodities, 42 percent were in trade, hotel, and restaurant businesses, and 11 percent were engaged in various activities, including street vending (Bezabih, 2000: 82; CSA, 1997: 8).

About 61 percent of the employment in Addis Ababa is in the informal sector. This proportion puts Addis Ababa among those cities with the highest employment in the informal sector in the world, surpassed only by Kinshasa (80 percent), Accra (70 percent), Lagos (69 percent), Mumbai (68 percent), Delhi (67 percent), and Abidjan (65 percent) (GTZ-UMAS, 2000: 41).

Of the informal activities, street vending comprises a widespread economic activity in Addis Ababa. The urban poor are typically involved in this activity. No one knows the number of street vendors in the city, but they contribute significantly to the city's employment. An integral part of the urban economy, they provide a wide variety of goods at bargain prices. One finds streets vendors and their markets in almost all the districts of the city selling goods (mostly food and personal accessories) and services without obtaining permits, giving receipts, or paying taxes. Some vendors operate at sidewalks, busy roadways and intersections where large numbers of people gather, or at bus stops, especially at the first and last stops. Others walk around the city selling goods or services without a fixed place from which to operate. By far, the largest numbers of street vendors are women who sell foodstuffs, such as bakery products, homemade bread, cooked cereals, fruits, and vegetables, to the city's poorest inhabitants. Hawking by women also includes charcoal and firewood selling, beer brewing, and kiosk operating. Most of these women toil day-in and day-out, but few are able to generate enough earning to meet their family's basic needs for food, shelter, and clothing. Women are the sole breadwinners for about 32 percent of the households in Addis Ababa (AACG, 1999: 10). Women are also major sources of income in many male-headed households, especially among the poor households.

Perhaps of all the people involved in informal sector livelihoods, street sidewalk vendors face more problems in the course of running their activities. Every business day poses a challenge to their survival because they do not have legal recognition. Some street-front shopkeepers do not want vendors to take up spots in front of their store for fear of competition. Others accuse vendors of blocking customers' way to their legally operating establishment. Shopkeepers or their guards thus constantly push vendors away from their storefronts. The security or police forces are the most menacing to vendors. As part of their daily routine, the police are duty-bound to keep vendors off sidewalks and streets. Vendors have to be on a constant lookout for the police because failing to do so may result in the loss of their merchandise should they be caught. So, every time vendors spot security officers they quickly pick up their goods and run away only to come back to their position as soon as the officers are out of sight. They are undeterred by police harassment, because giving up on their trade is tantamount to suicide: they can count on no other means of livelihood. The city government says it is harsh on vendors not only to protect the formal or established business sectors that pay the taxes but also to protect the public from consuming unregulated, unhealthy, or shoddy products. However, most pedestrians appreciate what vendors are doing because the latter provide relatively cheap and convenient services.

Urban agriculture is a growing informal sector activity in Addis Ababa. Many households living along some of the rivers that pass through the city choose this activity because it can offer employment, income, and food security for their families. The water that is extensively used for irrigating urban farms in Addis Ababa is the Little Akaki River. This river rises from the northwestern part of the Entoto mountain range and flows south passing through the western section of the city, cutting through the most densely populated districts of the city (Merkato, Ehil Berenda, Geja Sefer, and Lideta). Most of the city's large industrial establishments are located near or along this river. Small plots of farms are found on gentle slopes and level surfaces along the banks of this river starting from its middle course and extending all the way down to the town of Kaliti. Dense farms are especially found at four locations along the river: behind Amanuel Hospital, below the Addis Ababa Abattoir, across the National Alcohol and Liquors Distillery at Mekanisa, and behind the Kaliti Sewage Treatment Plant. All in all, about 160 hectares of land are irrigated to grow crops year-round. Farmers grow varieties of vegetables such as lettuce, cabbages, tomatoes, potatoes, red beets, carrots, green beans, and red peppers. Most sell their produce to major restaurants, hotels and supermarkets in the city. Others sell them in the local open market or groceries. While urban farming is valued, at the same time there is a growing concern about the health risks associated with the consumption of crops grown using water that is highly polluted with municipal sewages and untreated effluents discharged from industries located along city rivers (Berhe, et. al., 2000: 18).

Generally, households or individuals engaged in informal activities are handicapped by too many constraints. They lack almost all the essential ingredients to grow, such as access to bank loans, secured space for production and marketing, adequate legal protection, and essential physical infrastructure and services (Bezabih, 2000: 76). It is wrong to assume that all informal economies consist of hand-to-mouth economic activities performed by impoverished urban residents. Some incomes generated in the informal sectors are high enough to provide a reasonably decent living, and some are higher than formal incomes. Some workers in the informal sector make good incomes by selling their produce to formal sector industries (for example, urban farmers who supply vegetables to restaurants and groceries; women who supply foodstuffs to groceries). There are some large informal sector operations in the city, however, they hide their operations from city authorities to avoid paying taxes.

Because the informal sector plays an important role in Ethiopia's urban economy, urban governments should nurture it by allowing these businesses access to lines of credit, formal training, and more tolerant policies. In addition, urban governments should make sure that the informal sector pays taxes and that its workers abide by the law. With a large employment base, the informal sector has the potential to boost the urban economy, form human capital, generate a surplus, and efficiently use and recycle local resources while requiring little capital investment from the government. Because women are the majority in the informal sector (and often bear disproportionately the burdens of poverty, poor education, and lack of jobs), government recognition and aid to this sector will also improve the opportunities and advantages to women in the urban areas (Todaro, 1994: 253-257).

## The Urban Housing Problems

Serious physical deficiencies abound in almost all urban areas in the country, but no other problem is as daunting as the housing problem. Urban areas, large and small, are increasing in population at rates generally not less 3 percent a year and often up to 6-7 percent a year. The number of houses built, however, is far below the quantity required to house the growing urban population. For the most part increased densities and the partitioning of existing dwellings are accommodating urban growth. At least two-thirds of urban dwellers in Ethiopia (that is over 6 million people) live in housing that is overcrowded and of such poor quality, with inadequate provisions for water, sanitation, and waste collection, that their lives and their health are incessantly in danger.[11]

Until 1974 the Ethiopian urban land tenure system was highly skewed. In Addis Ababa, a few wealthy landowners dominated the urban land property markets. Almost all land (over 90 percent in 1966) in the city was in the hands of a few wealthy citizens. The monopoly over land caused disastrous disparities in wealth and hampered urban development. Landowners focused their attention on commercial endeavors, private mansions, and luxury apartments that catered to high-income groups. Land that was leased to lower and middle income groups was done so at exorbitant prices. The price of urban land soared because of uncontrolled speculation by urban estate monopolies. High land prices affected demand and constrained supply of housing. In addition, because the system favored land speculation, small landlords and most low-income groups were forced into situations of overcrowding as dwelling sizes were divided between more people.

By the 1960s the housing problem was serious enough to elicit government legislation. This legislation allowed for the purchase of residential dwellings through the Mortgage Company of Ethiopia and the Savings and Home Ownership Public Association. Despite this legislation less than 10 percent of urban households could have afforded to buy housing under the mortgage plans of the Association and less than 5 percent could have utilized the benefits of the Mortgage Company. Thus, most urban residents continued to suffer under landowner-tenant lease terms (Ministry of Public Works, 1967: 38-39, 55).[12] Over one-half of the urban housing in all major cities and towns remained rental units: 69 percent in Dire Dawa, 67 percent in Addis Ababa, 66 percent in Mekele, 62 percent in Harar, 61 percent in Bahir Dar, 60 percent in Bishoftu, 58 percent in Shashamene, and 56 percent in Dessie and Asela (Wolde Mariam, 1966).

## Urban Land Use Policy Under the Derg Regime

The establishment of the Derg regime in 1974 involved a restructuring of urban land policies to break the virtual monopoly of a few landowners. All residential land not occupied by the owner was nationalized and pre-existing tenant-landlord terms and contracts were annulled. Ownership was restricted to one house. The reform was specifically for the benefit of the tenants, especially those at the lower income levels. Houses rented below 300 Birr had rents reduced further, anywhere from 15 percent to 50 percent, with the highest reductions at the lowest scale. The regime promised to provide urban residents with credit options and lands for business and residential construction and to redress the gross disparities in wealth, income and services by reallocating resources amongst urban dwellers. Individuals in need of residential housing would be given up to 500 square meters of land free of charge under the directives of the Ministry of Public Works and Housing (the amount granted free of charge was reduced substantially in later years) (Gabre, 1994: 285). The reform was strongly supported by those who had not enjoyed secure tenancies and its popularity was high during its initial period.

The Kebele Urban Dwellers Associations (UDAs) were created to manage all houses renting up to 100 Birr per month, to collect property taxes and urban land rents, to fund social and welfare programs with the revenue rents generated, and to compensate, through monthly payments, the previous landowners whose sole source of income was renting houses (Desta, 1996: 35). Up to 15 percent of the revenue generated from rents was to be used by them in maintenance, repair and construction of residential houses. Houses with rents over 100 Birr were under the control of a government agency (the Agency for the Administration of Rental Houses—AARH). The agency was also responsible for the construction of rental houses. Neither the government agency nor the UDAs were able to meet the huge demand for rental houses, as their revenues were too low for regular maintenance and could not support new construction projects.

The Derg regime instituted various housing schemes to tackle urban housing problems. Individuals who wished to build their own houses were encouraged to do so by grants of 250 square meters of free land. They were also supplied with a housing plan. In addition, if a household satisfied the requirement of earning an income of at least 250 Birr per month, it was eligible to housing loans from the Housing and Savings Bank (HSB) at 7 percent annually. Another self-help housing scheme targeted income groups of 100 to 200 Birr per month. Individuals in this income bracket were encouraged to construct cooperatives, providing labor themselves while the Ministry of Urban Development and Housing (MUDH) provided loans (at 6 percent annual rate), technical support, supervision, and contractors. Households in the income bracket of 250 Birr and above were encouraged to partake in housing cooperatives of about twenty households. The HSB supported this group with loans at 9 percent interest and each household was given land up to a maximum of 250 square meters. Another scheme involved direct government participation in the construction of low-income housing on a limited basis (MUDH, 1983).

Despite the implementation of these different schemes, housing demand far outstripped housing supply. For example, in the years 1976-1981 the demand for houses reached 71,600 per year, whereas the supply averaged only a little over 6,000. Various factors contributed to the failure of these schemes. First, most of the population could not utilize home financing options because they did not fit the income eligibility criteria and because of the high rates of interest. For example, in 1978, 60 percent of urban households had monthly incomes less than 100 Birr, 80 percent below 200 Birr, and 86 percent below 300 Birr. Only 36 percent of the houses built between 1976 and 1981 were funded by loans from the HSB. One-half of the total cost of housing was incurred by bank financing during this six-year period. Government supplied houses only accounted for 13 percent of all housing in the same period. In addition, government housing schemes proved to be very expensive and catered not to low-income groups but to groups who were less in need. Urban dweller's associations built 30 percent of the units constructed under the other scheme. These constructions were the least expensive of all the units. Second, wages and salaries declined persistently after 1974. Drops in real income coupled with increased inflation did not stimulate investment in housing. The increase in prices of basic necessities and construction material was high because of declines in supply levels. Rapid urban population growth did not help matters either (Kebbede, 1992: 47-49).

Hence, under the Derg regime the housing situation in Ethiopia's urban centers became worse. Abolishment of the old landowner-tenant system eliminated land speculation, uncontrolled rents, and threats of evictions; however, various setbacks countered these advantages. Low-income groups were most affected by the crisis: they suffered from overcrowding, inadequate social services, and deteriorating environmental conditions. Most upper and middle-class income groups were not as badly affected, for they were able to build better housing or move to better neighborhoods if the need arose.[13] In addition, corruption plagued the urban land tenure system. Bureaucrats responsible for distributing free land meted out more land to favorites than to others. They indulged in illicit land sale practices that made them financially powerful. Their control over rental houses allowed them to overcharge on an already scarce commodity. In essence, these bureaucrats took on the vestiges of the old landowners and created an equally corrupt and exploitative class of 'renter landlords'—one with the right connection to be able to sell land on the black market, often the same plot more than once (Gabre, 1994: 287). Overall, during the Derg regime, the housing needs of the urban population, especially the low-income population, were met neither through the private real estate market, because there was none, nor through public housing programs, because there were few.

In sum, the seventeen years of the Derg rule were not kind to urban areas. Owing to bad management, corruption, ponderous bureaucracy, decline in capital expenditures, and misguided urban policies, urban areas had deteriorated greatly by the time the regime was removed from power in 1991. The new government inherited a city with problems of immense magnitude: inadequate water supplies and sanitation facilities, severe shortages of housing, dilapidation, critical infrastructural deficiencies, acute unemployment, extreme poverty, and an appalling decline in the quality of the urban environments, both built and natural.

## The Post-Derg Urban Land Use Policy

Article 40/3 of the Constitution of the post-Derg government states that land is the common property of the people of Ethiopia: 'The right to ownership of rural and urban land, as well as of all natural resources, is exclusively vested in the State and in the peoples of Ethiopia. Land is a common property of the Nations, Nationalities, and Peoples of Ethiopia and shall not be subjected to sale or to other means of exchange' (GTZ-Urban Management Advisory Services, 2000: 20). The state controls both rural and urban lands, as it did during the previous government. However, there is a big policy difference between the two governments in terms of how citizens acquire usufruct rights to land. During the Derg regime urban land was allocated to individuals and cooperatives free of charge for the building of owner-occupied houses. The sizes of land allocated varied with income but did not exceed 500 square meters at any time during the regime's tenure. The current government, on the other hand, terminated the free allocation of urban land and replaced it by the Urban Land Lease Holding Proclamation No. 80 (1993) which instituted a lease holding system of urban lands. This decree details several goals for efficient land use based on the principles of the free market. The objective of the new system of land allocation includes the creation of conditions that would improve the value of 'land use rights.' The system aims to find alternative ways to generate urban revenue and eliminate loopholes that have fueled the unjustified appropriation of gains. Other objectives include conscientious development of the urban landscape, an expansion in low-cost housing construction (using revenues generated from leasing land), and encouragement of private sector activities (Desta, 1996: 42).

The urban lease laws set the foundations for the public leasehold system. All urban land leases were to be established with the following rules in mind. First, leases were established on the basis of competitive tendering. Second, the duration of the lease should allow the lessee to recover his/her investment. Third, lease rights are transferable—fully or partially, depending on the specific lease. Fourth, leases can be renewed as long as property is not needed for public purposes. Fifth, the individual nature of leases can vary, but all leases must comply with zoning laws. Sixth, the government can terminate a lease if building laws are ignored, if the lessee is unable to meet financial burdens, and if the land is needed for public purposes (Council of Representatives, 1993; Desta, 1996: 42-44).

All urban lands, with the exception of those previously used for the construction of residential houses, are governed by the Proclamation. In addition, residential properties acquired through means other than inheritance are also subject to the lease policies. Urban land lease holding is approved if it is in accordance with urban land use patterns and has been through a competitive tendering. The purpose and use of land (for example, residential, commercial, industrial, educational, cultural, science and technology, or health and sport) and the level of urban centers determine the land lease terms and duration. The maximum duration allowed is 99 years for private residential dwellings at all levels of urban centers (Gabre, 1994: 290).

The Addis Ababa urban land lease holding regulations were issued in 1994 (Regulation No. 2). This regulation laid the foundation for minimum bid prices of land (as per zone and grade of land) and outlined provisions for lease transfer. The urban lands that were placed under the lease holding system included commercial lands, private residential lands, private residential lands that have been transferred via sale or as a gift to another person, private residential lands that are now utilized as commercial lands, lands used for the construction of apartment buildings that were previously used for private dwellings or commercial activities, and land to be used for rental houses (Council of Representatives, 1994; Gabre, 1994: 291-292). Urban lands that were held before the proclamation as private residences and private residences transferred through inheritance are not subjected to the lease holding law. However, by law the government can expropriate urban land and housing by giving due compensation. This eminent domain does not make city residents feel safe and secure about ownership of their property.

All urban lands are granted through a competitive public auction. Lands held prior to the proclamation are not subject to bidding, but are required to pay the lease rate determined by the tendering of nearby properties. Lease prices are based on five broadly defined land use categories and range between 6,000 Birr per square meter for first zone, grade one land, and 33.25 Birr per square meter for fifth zone, grade three land. Leases are set at 99 years for residences, 60 years for industrial land use, and 50 years for commercial properties. Lessee property rights are transferable via sales, pledges, or shares. Capital gains tax (30 percent) and an additional income gains tax (10 percent) are levied on all transactions. Lease payments require at least a 30 percent down payment; the remainder must be paid along with interest over a period of 10 to 30 years, depending on the intended use of the land (Gebre Giorgis, 2000; Desta, 1996: 46; Gabre, 1994: 292).

The regulation allows middle- and low-income individuals to lease a plot of land between 72 and 175 square meters for a period of 99 years upon lease payment of 0.5 Birr per square meter. Plots over 175 square meters can be obtained through the public tendering system by paying at least 30 percent of the cost at the time of acquisition. The remainder of the lease must be paid with interest over a period of 10-30 years. Urban land can be given free of charge and/or by bypassing the tender system to investments and land uses that the government encourages (for example, for social service institutions, low-cost housing, and other public facilities).

The regulation also encourages investment in rental units by reducing leasehold payments for those who build rental houses that fetch income not exceeding 800 Birr per month per unit. The reduction in cost of the lease is inversely proportionate to the rent charged on such housing units (Gabre, 1994: 293; Council of Representatives, 1994). For instance, an investor who anticipates to draw a monthly rental income of 100 Birr or less can write-off 90 percent of the lease price. Above a rental income of 100 Birr, lease reduction goes down by 5 percent for every 50 Birr additional income. In other words, rental income of up to 150 Birr per month brings down lease price by 85 percent; for 200 Birr income, lease reduction is 80 percent; for income of 600 Birr, the reduction is 30 percent, but for rental incomes between 601 Birr and 800 Birr, the relief is only 10 percent (Gebre Giorgis, 2000: 40). In spite of this, however, not many real estate developers have gone into rental housing construction. The reasons for this are quite obvious. First, the mortgage interest rate (at 10.5 percent) is too high. Second, taxes levied on rental houses are very high (up to 15 percent for rental income of 10,000 Birr per year; a flat rate of 45 percent for rental income of 50,000 Birr or more) (GTZ-UMAS, 2000: 40; Gebre Giorgis, 2000: 40). Third, capital gains tax (that is, tax on profit realized from selling a house) is 30 percent, a rate many real estate developers consider a rip-off. This last point is even more detrimental for first time homebuyers because sellers usually pass on the capital gains tax to buyers.

Rental housing units on leased property will not come cheap even if investors are willing to invest despite unfavorable lease terms (Gabre, 1994: 295). Rents will reflect the investors' cost of production (construction cost, lease payment, and taxes on property and income) and at least a 10-15 percent rate of return. All added cost would certainly hike up rental charges, making such rental units unaffordable for middle and low-income renters.

The public leasehold system has not produced tangible results in terms of alleviating the urban housing crisis, promoting equitable and efficient urban land use, improving the quality of the urban environment, or expanding the availability of basic urban amenities. The leasehold system's many restrictions discourage potential land developers from investing in real estate. Private real estate investors are not allowed to collect income in excess of the value paid at the time the lease was transferred. If they do, all additional proceeds ought to be transferred to the government. This particular restriction is counter-productive: it reduces incentives to invest, is difficult to enforce, and encourages discretionary behavior and under-reporting of prices. In essence, it affects the government's ability to generate surplus land value. In addition, the law is vague about provisions for government resuming leases, and how compensation will be paid to the lessee (Desta, 1996: 43-44). The absence of clear rules for end-of-lease arrangements discourages investors and is detrimental to the quality of urban development. As opposed to a freehold system, which would allow the landowner to implement long-term decisions, a lease system promotes insecurity and curbs the right to urban property ownership. If people cannot enforce their rights to land and housing, their incentives for investing in them will be considerably reduced. People build or renovate less if they think there is a risk that the government might take away or occupy what they have built. The effect of all of this is to reduce the aggregate investment in housing and other urban property improvements.

Lease prices are beyond the reach of most urban middle-and low-income people who aspire to have a home of their own. Additionally, leasehold contracts have little or no provision for the government to provide appropriate infrastructure and services for development. The absence of public provision of site-related amenities that accompany leased properties increases the cost of housing construction, thus making it doubly difficult for low-income households to invest in housing. In fact, the leasehold system deliberately favors well-to-do individuals or groups well connected to political organizations in power who invest primarily in non-residential buildings (mostly hotels, restaurants, office buildings, and merchandise shops) and luxury apartments and private mansions that only foreign residents and high-income persons could afford. Leasehold auctions are supposedly competitive and transparent, but winners are often predictable. There is no public access to sale records.

Exorbitant lease price is not the only problem that deters most prospective tenders. Even when investors are willing to pay a hefty lease price, the transfer of land is far from routine. Because of bureaucratic red tape and numerous administrative procedures the processes can take a year or two, at best. For instance, of the total investors that were granted investment permits in Addis Ababa between 1992 and 1998, only 29 percent succeeded in launching their projects (GTZ-UMAS, 2000: iii). The remaining investors dropped out due largely to frustration with the insurmountable municipal bureaucracy and corruption. The legal processes that are presently in place force individuals to navigate through substantial red tape to obtain property rights, requiring the payment of bribes and insider connections. Perhaps the most awful thing about the lease system is that the municipality can unilaterally call off signed and sealed lease agreements for no apparent reason and leave lessees in the cold. This has happened often enough to scare off many prospective real estate developers.

Severely limited access to capital is a major constraint to a potential house builder or buyer. Most people, especially low-income and middle-income households, cannot secure mortgage loans. The only public financial institution that provides mortgage loans is the Housing and Savings Bank, now renamed the Construction and Business Bank (CBB). During the previous regime mortgage interests were subsidized for low-income people. The then Housing and Savings Bank charged a 4.5 percent interest rate for cooperative housing builders and 7 percent for individual builders, but the current government did away with this subsidy. At present, the CBB charges 10.5 percent. Moreover, CBB's mortgage loans to housing cooperatives persistently declined from nearly 27 percent of total loans in 1992/1993 to about 8 percent in 1997/1998 (Gebre Giorgis, 2000: 26). Currently, the CBB and other public financial institutions favor large commercial ventures regarding loans.

Many factors influence an individual's ability to build a home—household income, cost of land and building materials, mortgage interest rate, and amount of floor space. In Addis Ababa, for example, the costs of building materials are quite high and keep rising. For instance, the price of a piece of corrugated iron sheet increased from a little over 10 Birr in 1989 to 39 Birr in 1992. The price of 100-kilograms of cement went up from 14 Birr in 1989 to 37 Birr in 1992 to 52 Birr in 1999. Imported building materials (for example, electrical goods and plumbing fixtures) reach consumers with heavy import taxes, making them unaffordable for most house builders (Gebre Girogis, 2000: 17-20). Thus, the cost of housing, even for the smallest shelter, is too expensive for most residents of the city. For example, a 60,000 Birr house with a floor space of 56 square meters (the minimum set by the city administration) would carry a monthly mortgage of 540 Birr (at 10.5 percent interest rate for 20 years). Assuming an individual is willing to pay a quarter of his/her monthly income for the mortgage to build such a house, that individual's monthly income ought to be 2,160 Birr. Only 4 percent of the residents in Addis Ababa earned 2,160 Birr or more (Gebre Giorgis, 2000: 14, 22) and 70 percent earned less than 200 Birr per month in 1995 (World Bank, 1995: 34).

## The Condition of Publicly-Controlled Housing

Much of the urban properties expropriated by the previous government (approximately one-half of all property in Addis Ababa alone) remain under the government or *Kebele* control. The government is reluctant to return these properties to their old owners or sell them to current occupants, even though it initially promised to do exactly that. The irony is that while the old owners are condemned to an impoverished life, the government allows the emergence of a new class of avaricious urban property owners. Urban real estate in Addis Ababa is falling into the hands of fewer and fewer individuals or political organizations, while the need for housing in the city increases.

Most of the *Kebele*-controlled houses are falling apart due to lack of vigilant maintenance. Often the housing stock is not structurally sound in the first place. Over four-fifths of the housing units in the Addis Ababa, for instance, are built with non-durable materials such as eucalyptus poles, mud and straw, wood and thatch, or bamboo with simple corrugated iron roofing. More than three-quarters of these housing units have earth/mud floors and stand on a very shallow foundation. This type of construction accounts for over 90 percent of the housing units rented by the *Kebeles* and most of these units have been falling into ruins for lack of maintenance for 25 years. Every so often, the mud structure must be given a new layer of mud mixed with straw to replace what torrential rains have washed away. Unfortunately, *Kebele* authorities have rarely taken responsibility for the task of repairs and maintenance of the housing units they controlled since the urban land and housing nationalization in 1975. The rains have peeled the mud cladding off the external walls of many of these housing units, causing the exposed timber frames to wear down. *Kebeles* are authorized by law to spend up to 15 percent of the collected rents for repairs and related expenses, but they rarely manage to provide more than superficial services: the total rental income is quite small and salaries, compensations to original owners, and corruption consumes a big portion of it. Two-thirds of the *Kebele* housing units are rented for 10 Birr or less. Rents for *Kebele*-controlled housing units have never been altered since they were set in the mid-1970s. Even though occupants are allowed to upkeep rented dwellings with municipal authorization, few seek an authorization to spend their hard-earned resources on a dwelling they do not own. Also, the inefficient and corrupt *Kebele* bureaucracy where one attains the required permit discourages those who may be willing to spend their own resources to improve their dwelling. The *Kebele* requires permits for even the slightest improvement (painting walls, for example) on the housing structure. Sooner or later, therefore, most run-down *Kebele*-controlled houses and buildings will have to be destroyed or they will crumble on their own. Unfortunately, the dilapidation of housing is not limited to *Kebele*-controlled dwellings. Disrepair in the private sector is not uncommon as many poor owner-occupiers find it difficult to keep their homes in good condition. With few extra resources at their command and the lack of access to loans for home improvements, many are unable to do anything other than a bare minimum of repairs.[14]

A 1995 survey of housing conditions by tenure showed that publicly owned properties were found to be a lot worse than owner-occupied properties. Nearly 56 percent of the public-owned housing units were characterized as fair or beyond repair while only 26 percent of owner-occupied units fell in the same category. Thirty-eight percent of the government-owned housing units that are occupied by public officials free of charge were also in poor condition and over 17 percent of them were found to be beyond repair (MWUD, 1997).[15] Most of these housing units were originally among the best dwellings in the city because they belonged to middle and upper class people.

The housing sector is a major productive sector of the urban economy, and, therefore, the private sector should be seen as a means to enhance urban economic growth and development. However, government and/or municipal governments have yet to provide loans, sites, and basic urban amenities to assist low-income groups in buying or building their own homes. The leasehold system has too many disabling constraints for the vast majority of the urban residents to own property. For many urban residents, it is too expensive. Furthermore, it discourages leaseholders from investing in the property they hold because of their uncertainty that the property and their investment gains will always be there for them to enjoy. The municipal government generates revenue through the sale of leases, taxes on property, and income, but little of this revenue is geared toward financing public services such as infrastructure, low-income housing, and environmental protection works. In the 1997/1998 fiscal year, land lease fees alone brought the Addis Ababa municipal government 29 percent of its total revenue (GTZ-UMAS, 2000: iii).

The failure of the housing supply to meet the demand has a serious economic consequence relating to the soaring house rents in the city. Families spend between one-third and one-half of their monthly income on rent of privately owned property. Actually, rental charges for urban housing are extremely uneven. Rents for *Kebele* owned houses are exceedingly low and do not reflect the real housing market at all. A four-room *Kebele* owned house, for instance, could go for as low as 30 Birr per month whereas a privately owned house with only two rooms may be rented for as high as 500 Birr or more. The demand for housing is so high that renters of *Kebele* owned houses illegally sublet a portion of their dwelling (often disguised as a lodging arrangement) for as much as two to three times the rent they pay for the whole house. In 1994, slightly more than half (50.8 percent) of the housing units in Addis Ababa were privately owned and the remaining units were under the control of *Kebeles* and the pubic housing agency (Environmental Protection Authority, 2000: 28).[16] Nearly three-fifths (56.8 percent) of the housing units were rental units. *Kebeles* controlled about 69 percent of these housing units, the AARH controlled 4.4 percent, and private holders owned the remaining 28.8 percent. Housing units occupied by government officials free of charge in Addis Ababa more than doubled between 1984 and 1994, from 12,500 to 29,000 units. The share of rent-free housing increased from 4.8 percent in 1979 to 7.9 in 1994 (Gebre Giorgis, 2000: 8), indicating a significant growth in the upper echelon of the government bureaucracy. Most of these high-ranking bureaucrats have their own homes, but they opt to lease them to foreign NGOs or embassies and enjoy the luxury of living in spacious government villas for free.

Housing in the city of Addis Ababa is inadequate and most of the houses are in deplorable condition, especially those located in the central parts of the city, lacking elementary amenities for healthy human survival. Many houses cannot protect residents from such natural environmental dangers as rain and extreme heat or cold due to punctured walls and leaking roofs. Overcrowding is a major problem. For example, according to the 1994 census, nearly a third of the housing units in the city had only one room with an average household size of 5.3 persons (United Nations criteria states that these should not exceed 2.4 persons per unit); about one-half had 2 to 3 rooms; and the remaining housing stock had four or more rooms. Over four-fifths of the housing units controlled by *Kebeles* had two rooms or less. Two or fewer persons occupied about 20 percent of the housing units; 3 to 5 persons occupied over 36 percent; and six or more persons occupied 44 percent. The consequence of high density per room is, of course, the availability of less floor space per person. Over three-quarters of the housing units in the city had a living space of 40 square meters or less (CSA, 1994). The health burden imposed on people living in dilapidated, overcrowded dwellings with inadequate provision of water, sanitation, and drainage is enormous. As the United Nations Center for Human Settlements (Habitat) studies show, for example, 'disease burdens from tuberculosis, most respiratory infections (including pneumonia, one of the largest causes of deaths worldwide), and intestinal worms are generally much increased by overcrowding' (UNCHS, 1996: xxviii). Overcrowded housing conditions also increase the risk of transmission for diseases such as influenza and meningococcal meningitis (WHO, 1992a).

In 1996, there were about 460,000 households in Addis Ababa, while there were only 238,000 residential housing units. That means, 222,000 households had no residential units of their own and had to double or triple up with others or had to resort to illegally build a shelter wherever they could find inconspicuous open space (Ayenew, 1999: 12). The poor are so desperate that they build shelters on hazardous land sites such as flood prone areas (on steep slopes) along the city's rivers. Flooding is a frequent phenomenon in the city resulting from the combined effects of steep slopes, sparse vegetation coverage or deforestation in the catchment areas (due largely to increasing demand for firewood and wood for construction that has been fuelled by the increasing urban population), and heavy torrential rains that occur during the summer season. Dwelling collapses are commonly reported during the summer season.

**Informal Settlements**

Illegal or informal settlements thrive in Addis Ababa. These are residential dwellings built on publicly owned land without permit or authorization from the municipal government. The proliferation of these houses—which are also known as *chereka bet*, meaning houses built by the light of the moon—is a response to the critical scarcity of shelter in the city. Most of these settlements are located in the southern, southwestern, and northeastern peripheries of the city on lands the 1986 Master Plan designated for new housing expansion, public parks, green areas, sport fields, freight depots, and industries. In 1988, there were 4,394 informal housing units in Lideta, Kotebe, Mekanisa, and Nifas Silk areas, accounting for 1.6 percent of the total housing stock in city. Of the 94,135 housing units built between 1984 and 1994, 14,135 units or 15.7 percent were provided by the informal sector. Such settlements have now expanded in other parts of the city such as along Jimma road, Ayer Tena, Furi Hana (at several locations along the new Ring Road between Jimma and Bishoftu roads), Kaliti (north and south of Worku Sefer, around the city's sewage treatment plant, and Saris area), Gourd Shola, CMC areas, Kotebe (the largest settlement on over 470 hectares of land), and Kara-Alo (along both sides of the Dessie road). All in all there were an estimated 60,000 housing units in these settlements in Addis Ababa in 2000, occupying about 4 percent of the city's land or about 2,000 hectares (Table 2.2). An estimated 300,000 people lived in these settlements (AAMPRO, 2001: 7-9).

Illegal settlements in Addis Ababa are not quite like the squatter settlements found in other urban places of the developing world that so much has been written about. Of course, both types of settlements are illegal, but their similarity ends there. While squatter settlements take over unoccupied land in an organized manner—often by violently confronting public authorities—illegal settlements or chereka bet occupants in Addis Ababa seize vacant land more or less on an individual basis—often with a tacit approval of dishonest public officials. Furthermore, chereka bet dwellings are structurally better than squatter settlements. As Tewodros Tigabu (1991: 32) describes them, the majority of the houses built in chereka bet settlements, 'are not made from cardboard, plastic sheets, [flattened] oil drums, and the like, but rather at best [they] equaled or [were] even better than the existing slum area houses of the city center. What we have are mostly well-built traditional houses constructed with mud, wood, and iron sheet roofing and interspersed with hollow concrete blocks, brick and stone houses, sometimes plastered, painted, and located in a comparatively clean environment. Density and overcrowding are very low relative to [the] congested [Addis Ababa] city center and to those squatter settlements in Mexico City, Sao Paulo, Calcutta, etc.' Moreover, squatter settlements lack the most basic infrastructure and services whereas informal settlements in Addis Ababa are far better serviced. Water and road services are available in most settlements, but electric and telephone services are available only to some settlements located

**Table 2.2  Informal Settlements in Addis Ababa: Location and Size**

| | Location | | Area (ha) |
|---|---|---|---|
| Area / district | Wereda | Kebele | |
| Repi | 24 | 16 | 228.13 |
| Jimma road on the right side of the road to Sebeta | 24 | 16 | 138.75 |
| Jimma road, the location same as above but smaller in size | 24 | 16 | 47.5 |
| The settlements around the Ayertena UDPO housing settlement | 24 | 15 | 115.63 |
| Furi Hana, three spots along the left and right side of the 1$^{st}$ phase ring road | 19 | 60 | 288.13 |
| Kaliti around the waste water treatment plant, settlements along the river | 27 | 11 | 62.5 |
| Kaliti, Worku Sefer | 19 | 59 | 95.00 |
| South of Worku Sefer, settlements along Bishoftu road | 27 | 11 | 38.75 |
| North of Worku Sefer, Sarris area | 17 | 20 | 80.00 |
| Gourd Shola, south of transport ministry workers residence along settlement | 28 | 04 | 117.5 |
| CMC area, settlement north of the special housing project houses | 28 | 01 | 62.50 |
| Meri at south and northern part of the road to Ayat housing project | 28 | 03 | 90.63 |
| Kotebe, the northern part of the development along the | 16 | 22 | 48.75 |
| Main road – the range extends to the whole settlement | 28 | 01 | 46.88 |
| including the last development on both sides of the | 28 | 02 | 81.25 |
| Dessie road | 28 | 03 | 462.50 |
| *Estimated total area* | | | 2004.4 |

Source: Addis Ababa Master Plan Revision Office (2001), *Housing Component: Improvement and Development Strategy*, Addis Ababa: Addis Ababa Master Plan Revision Office, pp. 8-9.

along main roads. Settlers access the services lawfully even though their settlements are built on illegally acquired land. The settlers would not have accessed services if it were not for the lack of coordination of actions between the municipality and other public utility corporations such as telephone and power. Utility corporations rarely require a customer to present title of ownership before they install services to a household (AAMPRO, 1999a: 25).

Illegal settlements in Addis Ababa are bound to stay. The city cannot evict them because there are too many of them and they continue to grow. This realization prompted the city government to declare its intention in April 2000 to give legal tenure to informal settlements. But not all informal settlements are eligible to apply for legal title. Only those that existed before regulation 47/1974 and those holdings that were acquired via local authorities and peasant associations are covered under the new regulation. That leaves out more than 40 percent of the illegal settlements in the city (AAMPRO, 2001: 11). The municipality does not have a clear plan for the remaining illegal settlements. At any rate, this new regulation may have sent the wrong message to potential illegal settlers. It is not far-fetched to imagine that some people might go ahead and build shelters on an illegally acquired piece of land hoping that they too will in the future be legally entitled to stay for good.[17]

Poor people will not opt to build dwellings on illegally occupied land if they can afford legal plots of land. Perhaps land is the biggest failure in the development of housing in urban areas. The government has failed to stimulate a supply of sufficient, affordable, and serviced land (UNCHS, 1990: 32). Its regulations for land use and supply have caused land shortages. In most cases, governments in developing nations adopt such inefficient use land policies to garner political and economic gains (UNECA, 1996: 71).

While most people in Addis Ababa live in overcrowded houses, highly congested neighborhoods, and illegal settlements, there are also many without a shelter worth the name. Tens of thousands of people in Addis Ababa have no home and they sleep outside under the cover of salvaged materials such as cardboards, canvas, or plastic sheets set against fences or walls of public buildings (Gebre Girogis, 2000: 6). Others sleep on the verandas of commercial establishments. Many street-children among this homeless population lead lives with exposure to all kinds of dangers: for example, violence, sexual exploitation, vermin, pollutants, and exposure to extremes of weather. There are an estimated 40,000 homeless children in Addis Ababa alone and their number escalates at the rate of 5 percent per year (UNICEF, 1996).[18] Accurate information on homelessness in other major urban areas is too scanty to know if this problem is growing, but given the huge deficit in housing supplies and acute urban poverty it probably is.[19]

**New Real Estate Developments**

In spite of the dilapidation of much of the old city center and the acute shortage of housing to accommodate the city's growing resident population, no one would doubt that the government's liberalization policy has been too good for investors with big capital, including foreign as well as domestic investors. In the last ten years or so, Addis Ababa has witnessed unprecedented big-ticket real estate developments (Table 2.3). The biggest real estate development is the just completed Sheraton Addis Hotel, considered to be the largest, fanciest, and most expensive hotel in the whole of Africa. The hotel was built by bulldozing 300 slum housing units (Assefa, 1998: 1) that were located on the western slope of the old Grand Palace (the *Ghibi*). Other medium-to-large size hotels with much lesser names are popping up along some of the major roads in the city. Quite a few completed and ongoing commercial buildings are covering the city's skyline: Mohammed International Development, Research, and Organization Companies (MIDROC) office buildings across from the old soccer stadium (considered to be the tallest building in the city), Dembel (along the Bole Road), and Sam International (near Mexico Square). A number of big real estate investors—such as the Ayat, Safe, and Addis Real Estate—are involved in the construction of residential housings exclusively geared to meet the housing needs of people in high income brackets. Quite a few expensive private homes (with price-tags of at least half a million Birr) are also being built at choice locations.

Low cost housing is also being built at Asco, Ayer Tena, Ferensay, Kotebe, and Akaki (AAMPRO, 2000a: 10). All these real estate developments are good for the city's economy; they create jobs and the municipal leadership has the right to claim credit. But there can be no argument with the fact that these development efforts alone cannot make a dent in the acute housing shortage.

**Urban Transportation**

Ethiopia considerably lags far behind other developing nations in the provision of adequate transportation for its people, including road, rail, and water modes of transport for movement of people and goods to and from various places in the country. With only 31,571 kilometers of all weather roads, of which just 12 percent (3,788 kilometers) are paved (World Bank, 2003: 290), and only one passenger car per 1,000 people in 2000 (World bank, 2003: 164), the country is far from meeting the transportation needs of the majority of its people, especially those in the rural areas. The near absence of feeder roads is a major physical barrier for the movement of people and goods between regions and between urban areas and their hinterlands.[20] An estimated three-quarters of the rural population is more than half a day's walking distance from the nearest all-weather road (EPA, 2001: 126). In 1981/1982, there were less than 60,000 motor vehicles in the whole nation; that number grew to a little over 90,000 in 1995/1996 of which well over three-quarters operated in Addis Ababa. The overall number of motor vehicles grew by more than 50 percent over the last ten years (Tadesse, et. al., 2000: 39). Urban areas, especially the larger ones, are increasingly becoming motorized, but insufficient infrastructure and maintenance is handicapping their efficiency.

Cities cannot effectively function without transportation networks that move people, goods, and services promptly and efficiently. Transport has probably more influence on the pattern of land use than any other infrastructure facility in urban areas. Commerce and industry depend on transportation to carry out their daily activities. City dwellers move from one place to another for one reason or another. People and goods move in and out of cities daily. A large number of motor vehicles move between homes and places of work, market centers, and business areas. All these movements are bound to produce intense traffic flows.

Such intra-urban mobility can produce serious traffic problems in a large city like Addis Ababa where infrastructure development and maintenance is rudimentary at best. Most of the roads in Addis Ababa were built several decades ago for slow vehicles and pedestrians and are not adequate to cope with the needs of modern transport. The city has a mere 1,326-kilometer long road, but only 400 kilometers (30 percent) of this is asphalted (Tadesse, et. al.: 2000: 28) and less than half is fitted with adequate drainage system for storm water (Malifu, 2001: 145). The layout of the city's road network is poorly designed. Nearly three-quarters of the roads are narrow, only 6-8 meters wide (below the standard of 9 meters) (Tadesse, et. al.: 2000: 28). Many roads have sharp, dangerous curves; many go over bridges (for example, over the Akaki, Bambis, Qabbana, Ginfile, Afincho Ber, Ras Mekonen bridges). Five to six roads converge into major squares or intersections in the city making traffic flow extremely slow (for example, six busy roads merge into Mexico Square, Abnet Hotel, Kolfe Mesalemia, and Gotera junctions while 5 roads merge into Megenagna Square, Ourael junction, and Teklehaimanot Square). Sidewalks for pedestrians are also grossly absent; only one-quarter of the asphalted roads have sidewalks (Tadesse, et. al.: 2000: 55). Hence, cars, buses, taxis, motorcycles, and trucks must compete for space with pedestrians, donkeys, and domestic animals.

Consequently, problems of traffic congestion in the city's main arteries, increasing numbers of traffic accidents, and insufficient off-street parking facilities continue to grow. In Addis Ababa, the journey to work is becoming long, tedious, nerve-racking, and expensive. All these problems, coupled with the lack of adequate traffic signs and signals and the laxity in enforcing traffic controls and regulations complicate the city's transportation problems.

The urban traffic quandary is not only a problem of congestion but also one of inadequate mass-transit vehicles and inefficient planning and management. For a city with a population of nearly 3 million, the Anbessa City Bus has only a fleet of about 400 (that is, one bus for every 7,500 persons). The city's mass transportation has not grown commensurate to the increase in resident population.

**Table 2.3  Major Ongoing Projects in Addis Ababa: Real Estate and Other Infrastructure**

| Project Type | Projects under construction or already financed | Projects under study | Projects at proposal stage |
|---|---|---|---|
| Road network | North regional road Kajima project; Addis Ababa-Awasa regional road; First and second phases of the Ring Road construction | West regional road (Ambo Road); Jimma Road; East regional road (Dessie Road); third phase of the Ring Road | Construction and upgrading of pedestrian roads |
| Terminals | Additional runway and passenger terminal building (Bole Airport); regional bus terminal | Akaki freight terminal | |
| Sewerage | Liquid waste treatment plant | Liquid waste line network | |
| Drainage | Relocation project of flood control and prevention | Flood control and prevention | Artificial lake (for flood control) |
| Tele-communication | Arada, Yeka, and Gergi area telephone line installation | Renewal project of old telephone lines in the City | |
| Water supply | Akaki ground water project | Addis Ababa water supply project stage III; construction of Sibilu Dam | |
| Green framework | Entoto reforestation project (Ethiopian heritage trust) | | |
| City center redevelopment projects | ECA area redevelopment project; Sheraton area redevelopment projects | Merkato redevelopment and upgrading project; Adorna-Alpha project in south Merkato | |
| Important building complexes | MIDROC office building next to stadium; Sheraton extension; Dembel (Bole Road); Sam International (Mexico square) | MIDROC city center twin towers in Piazza; Awash Bank and Insurance office building (Bole Road and Tewodros square) | |

| Main facilities | Private General hospital in Keranyo; Selam General hospital in Mekanisa; Samson Sport Center | St. George club stadium in Keranyo for 60,000 spectators; Muslim festive area in Keranyo; American University in Kotebe | Olympic-sized stadium at Gergi |
| --- | --- | --- | --- |
| Housing | Real estate project in Kotebe (Ayat Real estate); low cost housing construction in Asco, Ayertena, Ferensay, Kotebe, and Akaki; private residential constructions in the main expansion areas | Bole Lemi detail study; Lideta Real Estate project (Huda Real Estate); Sheraton area relocation study | |

Source: Addis Ababa Master Plan Revision Office (2000a), *Addis Ababa Revised Master Plan Proposal: Draft Summary*, Addis Ababa: Addis Ababa Master Plan Revision Office, p. 10.

In the last ten years alone, the volume of passengers increased by 86 percent while the bus fleet grew by less than 20 percent (Tadesse, et. al.: 2000: 45). Moreover, no more than 290 of the buses provide scheduled service at a given time during the day due to frequent breakdowns and bus schedules are dreadfully unreliable and uncoordinated. Buses are frequently overloaded with passengers. On the average passengers unnecessarily spend more than 30 hours per month waiting for buses (*Tobia*, 19 October 2001). It is in response to this severe inadequacy that privately owned passenger vans or mini-buses (that accommodate 12-16 riders) and taxis (with capacities of 5 or less riders) have burgeoned in the city in recent years, growing by over 10 percent per year in the last 5 years alone (Tadesse, et. al.: 2000: 43). Nevertheless, the city's transportation services suffer from huge capacity problems at peak hours.

A transportation network system that is lacking timely maintenance and upgrading can undermine the city's economic productivity. Time wasted on dilapidated roads can exact losses to businesses, large and small. It hastens the wear and tear of motor vehicles, jacking up repair costs to owners. Congested roads slow the timely delivery of goods and services and increase the cost of operation (World Bank, 1994: 25-26). Congestion also causes higher fuel consumption and increases airborne lead pollution because many city vehicles are old and use leaded gasoline. Data is scarce on the level of lead emission in the city, but it is well known that lead is highly dangerous to human health. Addis Ababa consumes about 63 percent of the gasoline and 23 percent of the diesel fuel used in the country (Tadesse, et. al., 2000: 49). Apart from inadequate public transport and congestion, other problems of the road system in Addis Ababa include poor maintenance and traffic accidents. Most of the city roads are in complete disrepair. Potholes go unrepaired far too long. Such delays usually increase the cost of maintenance. Many neighborhood roads that were once asphalted have now turned into rock-strewn roads. Most taxi drivers avoid driving on these roads no matter how much money people are willing to pay them. Traffic accidents are a major cause of death and injury in the city. Addis Ababa had 59 accidents per 1,000 vehicles in 1993 and 109 in 1998. By comparison, in 1991, Mexico City had 4 accidents per 1,000 vehicles; Bangkok, 16; and London, 3; Rome and Buenos Aires had 9 and 14, respectively, in 1993. Ethiopia is among the top six Africa nations for road traffic death rates per 10,000 vehicles—the others being Ghana, Rwanda, Lesotho, Kenya and Cameroon (Jacob, 1997).[21] In Addis Ababa, there were 7,345 motor vehicle accidents in 1998/99, of which 280 resulted in fatality (CSA, 2001b: 197). Damages to property because of traffic accidents are quite high as well, estimated at 12 million Birr a year (Tadesse, et. al.: 2000:64).

Pedestrians make up more than 90 percent of the city's traffic accidents while drivers and passengers share the rest in almost equal proportion. Pedestrians also suffer over four-fifths of the injuries. Since most people in the city cannot afford public transportation (even with the Anbessa City Bus subsidy of about 0.26 cents/person/trip), they have to commute back and forth to work on foot. In fact, over 70 percent of the work commuting in the city is done on foot. In the absence of sidewalks these commuters have to share the roads with cars and pack animals (Tadesse, et. al.: 2000: 35, 45).[22] As they walk, they obstruct traffic and expose themselves to a high risk of accident. Street vendors occupy the available sidewalk, forcing pedestrians to walk on the streets. Pedestrians contribute to their own death or injury. It is quite bewildering to see pedestrians not particularly concerned about their own safety. They seem to feel a sense of entitlement to equal share of the street with motor vehicles. What is even more perplexing is their disregard for traffic safety rules and their complete trust in the drivers for their own safety.

The primary causes of traffic accidents are speed, reckless maneuvering, encroaching into opposite lanes, failure to yield when it is appropriate to do so, and disregard for traffic signs and rules. Most drivers feel that they own the roads. Drivers get away with almost all traffic violations, because corrupt traffic police take bribes instead of issuing tickets. Poor road conditions, the absence of adequate traffic management and control facilities, the absence of sidewalks, and poorly maintained and not-roadworthy motor vehicles also contribute to traffic deaths and injuries. For a city with thousands of major road intersections, there are fewer than three dozens traffic lights. A third of these do not function at one time or another. The ones that function have problems of synchronization and often need intervention by traffic police officers to manage traffic flows, especially at peak hours. Potholes and open manholes also add to traffic safety problems in the city. Many traffic accidents occur as motorists attempt to dodge potholes.

Traffic congestion is expected to get worse in Addis Ababa with increasing development in all peripheral areas around the city except in the north. The Addis Ababa Master Plan Revision Office (1999b: 23; 2000a: 10) projects increasing congestion in the following routes. First, the east-west route from the Army Hospital to Megenagna will become more congested because of extensive developments in Kotebe and Qaranyo areas. A real estate project (the Ayat) is a major ongoing development in Kotebe, while in Qaranyo the construction of a stadium for the St. George Soccer Club that is expected to accommodate over 60,000 spectators and a dedicated space for Muslim holiday festivities are under consideration. Second, the Bole Road, which already carries dense traffic throughout the day, is expected to get worse due to developments along the road itself and in Bulbula. Third, the Lafto-Mexico Square Road will have dense traffic due to extensive developments in Mekanisa area and Lafto, one of the fastest growing peripheral communities in the city. Fourth, traffic density on the Bishoftu Road, which is now the busiest route in the city, is expected to increase due to the flow of people and goods between Addis Ababa and the eastern, southeastern, and southern regions of the country. This road is the only entry route to the city for people and goods coming from these regions. Many of the industries and warehouses in the city are also located along this route. Fifth, the Merkato area, which is the largest commercial center in the city, is expected to see further business expansion, but the traffic congestion at the present time defies description. At any given time of the day the street here are the most densely populated streets in the city in terms of the volumes of pack animals, people, and motor vehicles. Finally, traffic volumes in the old city center (Piazza) are projected to increase due to a planned MIDROC twin towers construction on lots that have been vacant for several decades. The existing narrow, winding roads in the center are not expected to handle the growing traffic. The absence of parking lots in the area will exacerbate the situation as well.

The idea for the ongoing construction of a thirty-three kilometer Ring Road or Expressway around the city was to ameliorate the problem of traffic congestion inside the city. However, this is not going to do the job until artery roads that merge into the Ring Road are built. At present motorists hardly use the completed 11-kilometer portion of the Ring Road from Kaliti to Ayer Tena because it has no exits that connect with roads into the city. Moreover, it is not a real Ring Road because much of the road passes through densely settled residential neighborhoods and commercial and industrial land uses. The areas along the Ring Road from Ayer Tena to Wingate, for example, are dominated by commercial and residential land uses. The area along the Ring Road from Kaliti to Bole is dominated by residential use while the Bole-Megenagna section is dominated by manufacturing and service activities. Pedestrian movements are hampered along all these sections of the Ring Road. Pedestrians will not be able to cross over to the opposite sides of the Ring Road, without the risk of getting hit by a motor vehicle, when traffic density picks up after the completion of exit routes. That is because the overpasses that were supposed to facilitate pedestrian movements across the Ring Road are located several kilometers apart.

**Urban Governance and Management**

The Ethiopian constitution recognizes the basic human right to adequate housing as articulated in the Universal Declaration of Human Rights and in the United Nations Global Strategy for Shelter to the Year 2000: 'adequate, safe and healthy shelter is essential to a person's physical, psychological, social, and economic well-being and is a fundamental component of sustainable development. It means adequate privacy, adequate space, adequate security, structural stability and durability, adequate lighting and ventilation, adequate basic infrastructure, water, sanitation and waste management, and adequate location with regard to work and basic facilities, all at an affordable cost' (UNECA, 1996: 143).[23] However, over four-fifths of the people in Ethiopia do not have access to a safe and healthy dwelling in both urban and rural areas. The provision of adequate housing is currently the single largest problem for urban areas. In Addis Ababa alone, the housing need for the last two decades was estimated at 77,600 units annually while the actual supply has been on the order of 6,000 (Region 14 Administration, 1997: 1). This means, therefore, that unless urgent measures are taken, urban dwellers will continue to live in squalid, unsafe environments where they confront a host of threats to their health and security.

Many attribute urban housing and other basic service problems to rapid population growth. This simplistic explanation overlooks the fundamental causes of the persistent deterioration of urban life and environment. The deficits in the provision for housing, piped water, sanitation, drains, roads, health facilities, and other forms of infrastructure and service provisions in Addis Ababa and other urban areas are the products of failed government policies, the absence of competent and accountable urban governance, the complete disenfranchisement of citizens, 'corruption, inappropriate regulation, dysfunctional land markets, unresponsive financial systems,' insecure land ownership, and 'fundamental lack of political will' (World Bank, 2000a: 1). Urban population growth, rapid or otherwise, has very little to do with the lack of economic and social development in urban Ethiopia.

One cannot help but reach the overall conclusion that the condition of urban housing is appalling and government urban policy is largely to blame. The government has thus far been neither an enabler nor a provider of urban housing, especially with regard to low-cost housing. Fundamental changes in urban land and housing policy are needed to solve the housing problem. Policies ought to ensure 'competitive but regulated market in land, housing finance, and building materials and to remove unnecessary bureaucratic constraints on the different stages of housing production. This includes ensuring that there is a ready supply of land for housing in urban areas with the price of legal housing plots with basic services kept as low as possible' (UNCHS, 1996: xxix). Additionally, the government ought to encourage public-private cooperation to facilitate community-based self-help solutions, promote the use of low-cost material and labor-intensive construction techniques, develop local designs and construction technology, provide urban land use planning and infrastructure, and create security of ownership and occupancy (UNDP, 1998: 89). The need for the latter measure is crucial. In this regard, laws governing public ownership of urban land and housing must be rescinded in favor of private ownership. Ownership of urban land and housing motivates people to improve and maintain their dwellings and neighborhoods. The building and improvement of housing is 'a universal activity—and people everywhere demonstrate ingenuity and creativity in doing it.' If people are provided security of tenure and occupancy, they are 'willing to invest great efforts in their homes—adding to savings and investment while improving living standard. Most of this can be done without direct government support—but governments, especially local authorities, have the obligation to ensure an enabling environment to release this creativity' (UNDP, Ibid.).

If the policy measures discussed above are to make any headway, there will have to be genuine political will and good urban governance in the country. Probably one of the most disturbing problems of urbanization in Ethiopia is the lack of good urban governance. Urban areas, like any social or economic organization, require qualified, enlightened, experienced, and innovative personnel to function effectively. Unfortunately, urban municipalities in Ethiopia do not have the required urban managerial and administrative expertise. The urban management problems are further worsened by the high rate of official corruption and mismanagement of public money among many personnel in charge of city administration.

Addis Ababa's current state of affairs presents a good illustration of urban governance and management problems. The city suffers from weak institutional capacities, a serious absence of popular trust, and a pitiable social and economic service record. It needs a government that feels accountable to its citizens and that encourages their participation in municipal issues. Meheret Ayenew, in his excellent piece on the state of municipal governance and management in Addis Ababa, persuasively argues that most of the city's problems stem from the lack of astute and responsible municipal leadership that is accountable to the residents. He says, first, that 'there is little awareness on the part of the municipality's leadership of the core functions and responsibilities that a modern city administration should undertake.' Secondly, he says, 'the current governance and management structure of the city is highly politicized.... [M]ost of the people in positions of authority within the administration of the city got their positions not by their merit or through professional competence but because of their political allegiance' to the party in power (Ayenew, 1999: 1-2).

The municipal leadership is not accountable to the people it governs, but to the government above. Ninety-six council members, elected for five years, administer the city. While the full council draft municipal laws and formulate policies, a 15-member executive committee manages the city's daily businesses. The chairperson of the council and the executive committee is the city's top administrator or governor. Of the three branches, the governor has the least power. The position is mostly ceremonial: the governor represents the city but has little influence over its operation. Power rests on the executive committee. The committee safeguards the laws of the federal government and the city council, oversees city government agencies, submits the annual budget, and prepares and executes development plans. By law, the city council is accountable to the electorate and to the chief executive of the federal government, the Prime Minister. The city Charter grants the Prime Minister the right when necessary to disband the city council and to order a new election within a year (MWUD, 1999: 65). Under this circumstance it is not hard to imagine that accountability to the Prime Minister would have stronger weight. In addition, almost all the municipal council members are also members of the ruling party in political power (Ayenew, 1999: 6, 9). The Federal Constitution guarantees the citizens of Addis Ababa a 'full-measure of self-government.' However, that is in not the case in reality.

There can be no argument with the fact that the Addis Ababa municipal leadership has failed to address the social, economic, and service needs of the city's residents. Poverty is widespread. About one-half the population of Addis Ababa earns less than what is needed to purchase adequate food for basic subsistence (Thomas and Taylor, 2000).

Unemployment and underemployment are rampant. Vital urban amenities are in short supply. Acute shortage of housing has resulted in overcrowding, homelessness, and neighborhood congestion. The physical infrastructure has fallen apart beyond recognition. Judged by all urban measures, the city may be characterized as one of the most dilapidated and environmentally degraded cities in the world. The current municipal leadership recognizes all these problems; it sees and lives them daily, but does little to solve them. Without effective and efficient municipal governance, the problems of the city will multiply.

The current municipal leadership cannot meet the challenges of the time: to plan for and manage city growth. The corrupt, unenlightened leadership lacks a vision for the future of the city. It cannot even collect taxes. During the second half of the 1990s, the municipality failed to collect over 4 billion Birr. This failure stems largely from the lack of adequate trained personnel to assess and collect taxes as well as an outdated administrative and financial system that is disreputably 'inefficient, inflexible and time consuming.' Corruption is also a major factor contributing to revenue underperformance. Tax managers, assessors, and collectors routinely under-assess and/or under-collect taxes or encourage outright evasion in return for a bribe in cash or material rewards (Ayenew, 1999: 26). Throughout the 1990s, the city experienced deficits and had to depend on the largesse of the federal government.

As Ayenew (1999: 30) has aptly observed, 'the first essential step to improve the governance and management of the city of Addis Ababa is to give the city genuine autonomy and make its administration responsive to the citizenry.' At present a single political party dominates the city administration. The current municipal administrators pay no political penalty whatsoever for their failure to adequately provide the most basic services, for their inability to maintain law and order, for being corrupt and utterly incompetent to do the job they are paid for, and for paying no heed to citizens' legitimate complaints. In the absence of competing political parties citizens cannot vote-in the leaders they like and vote-out those they do not.

Urban areas in Ethiopia need sufficiently public-spirited administrators who can measure success in terms of the quality and quantity of the services they provide to the residents who elect them. This type of dedicated leadership can help ameliorate some of the problems that pertain to urban development and governance in the country. The day-to-day running of the urban areas must be handled by well trained, experienced, and professional administrative and planning personnel who are interested not solely in their personal gains, but in the economic and social well being of the city. Addis Ababa desperately needs a truly democratic representation in the city hall and a leadership that facilitates participatory processes by which city residents could design the city development strategy to reflect their vision and aspiration for the city, monitor and evaluate economic development progress, and identify problems and suggest possible actions to be taken to solve them.

One cannot say enough about the indispensability of citizens' participation in urban development. One of the most glaring weaknesses of urban governments in Ethiopia is their reluctance to recognize and support the efforts of individual citizens, households, community based organizations, self-help groups, voluntary organizations, and small-scale private enterprises. As HABITAT has pertinently observed, 'many of the most effective and innovative initiatives to improve housing conditions among low-income groups have come from local NGOs or community organizations, including women's groups.' However, in most urban areas of the developing world like Ethiopia, 'the individual, household and community efforts that help build cities and developed services have long been ignored by governments, banks, and aid agencies and are often constrained by unnecessary government regulations.' If urban governments are willing to enhance the role of their citizens, 'what can be achieved by supporting the efforts of several hundred community organizations in a single city can vastly outweigh what any single government agency can do by itself' (UNCHS, 1996: xxx).

The absence of effective urban management, acute shortage of housing, inadequate infrastructure and basic amenities, and unplanned urban growth are, of course, not limited to Addis Ababa. Urban areas in Ethiopia share, by and large, similar problems, but certainly to varying degrees. An excellent 1999 study of the Amhara Regional State by a team with members from federal and state governments and expert consultants from the private sector captures the conditions of urban areas in other national regional states. The team based its study on interviews with urban municipal authorities and rank-and-file workers, district (*Wereda*) and regional council members, and a diverse group of resident populations. The subject cities included Gondar (population, 156,087), Dessie (135,529), Bahir Dar (134,062), Kombolcha (55,122), Debark (20,026), and

Addis Zemen (19,912). Below is a summary of the team's findings on a number of key urban issues: *the legal identity, governance, and management of cities; personnel management; urban revenues; urban planning; intergovernmental relations; service delivery; community relations; and vision, mission, and growth of cities* (MWUD, 1999).

*Legal identity, governance, and management of cities* Under the new Federal Constitution that provides for a decentralized form of local governments, the national regional states oversee the concerns of local governments including municipal affairs. The Amhara Regional State constitution does not accord cities self-rule. Therefore, urban municipalities are not autonomous entities and do not have their own elected council members that are accountable to the residents that elected them. Not a chief city executive and a band of professionals below him, but rather a committee of council members manage the cities. Party loyalists fill municipal political positions whether or not they merit them. Thus, not achievement but political connections determine municipal offices. City management is poorly structured. Technical and non-technical responsibilities are confused; line services and support services are entangled. Most senior staff do not possess the required technical knowledge to carry out functions assigned under them. Department heads are unable to make decisions on their own for lack of entrustment. Even the most ordinary issues that could be easily be handled by lower level personnel have to be referred to the top executives (MWUD, 1999: 120-25).

In underscoring the ineffectiveness and the less than democratic nature of urban governance, the team writes:

'Each city had a system of governance structured in a pattern set by the State. There were no choices as to how one city or town was to govern itself. The system of governance varied according to the grading given to it. Thus, Bahir Dar, the State Capital, had special zone governance with its council being constituted of *Kebele* representatives and with the election of the members of its Executive Committee being approved by the State. The *Wereda* towns of Gondar, Dessie, and Kombolcha were overseen by *Wereda* councils and were run by executive committees whose members were selected by the respective councils. Whereas the municipal towns of Debark and Addis Zemen were run like sector bureaus within their respective *Wereda* having no council of their own save the representatives of their *kebeles* in the *Wereda* council. None of the cities or towns had at large elections and none had mayors as such. The special zone city and the *Wereda* cities had chairmen who also chaired their respective executive committees. The executive committees essentially played the role of a chief executive officer with their chairperson being essentially the linking pin more than anything else. Again here …none of the chairmen or members of the executive committee were elected at large.'

'…[T]he governance in all cities…did not restrict itself to just policy-making. It was also handling the management. This dual role had two obvious side effects. First, the governance could not pay enough attention to the concerns that pertain to urban growth and development. Governance had a dimmed vision and its creativity about change within cities was very much limited. Second, governance seemed to have pre-empted the place of management muzzling proper roles in this regard and obviating the effective place that professionals could have in urban development. For all that could be observed, the interference in management had a poor showing in terms of efficiency, effectiveness and change' (MWUD, 1999: 121).

The report indicates that municipal officials recognized their inability to effectively serve the public because of poorly defined roles and responsibilities as well as a lack of investment, vision, and long term planning. When asked how to improve city governance, officials offered several solutions. City functions need to be identified and assigned to separate sectors within the city's government. Management of day-to-day functions and public services should be the responsibility of an appointed city manager whose activities are entirely separate from the city's political arena. The city council and executive committee should focus only on policy making and professionals should be left to implement the policies. A city mayor should replace the chairman, taking on a leadership role in the government to improve efficiency. Council members should be trained in urban leadership because their lack of experience has served as a major roadblock to productivity in the past. The number of council representatives should reflect equal proportions of urban and rural representatives so that all communities within the city can be represented in an equal manner and have their needs met. Lastly, officials noted that the government should be held accountable to one authority. Currently, higher levels of government including regional officials regulate city government, while the ever-present public goes largely ignored. Many officials agree that city governance should be self-regulating with a primary focus on public approval through feedback from their constituents. For that to happen officials emphasize the need for the creation of a municipal government composed of an elected city council that is accountable to the public and responsive to the public's needs.

*Personnel management*  Officials complain because city governments are not independent entities and that they are required to consult with state agencies of the Civil Service Commission and the Bureau of Works and Urban Development when hiring for municipal government positions. This makes the process time-consuming and expensive. Because state agencies are so removed from the communities they oversee, their guidelines fail to account for urban realities. Approved positions do not always reflect municipal needs; some departments are thus overstaffed while others are understaffed and officials do not have the power to reassign jobs in response to municipal needs. Officials complain that they cannot attract or keep qualified employees as a result of non-negotiable low salaries, inadequate benefits, and poor working conditions. The system does not reward efficient and ingenious employees. State mandates also make it difficult for municipalities to fire corrupt and incompetent employees. Officials recommend that cities be permitted to develop their own guidelines just as private companies do. From the city residents' point of view, however, many municipal personnel are totally inept, corrupt, and abusive. City residents suggest that allowing municipalities to fire employees at their discretion and promote or reward exceptional employees are important steps to improve municipal government.

*Urban revenues* The lack of fiscal autonomy prevents municipal governments from revising tariff rates and taking advantage of new revenue sources to meet the increasing demand services to finance the 'long overdue capital expenditure' (MWUD, 1999: 96). In many cities, tariffs have remained static while urban populations continue to grow. City revenue has thus not increased to meet the increased city expenditures. Officials agree that taxes should be higher and more frequent in municipalities. Increasing property taxes would be inadequate because too many public corporations are currently non-taxable including public corporations such as telecommunications, power, Commercial Bank, and federal state, and zonal government properties. Urban revenue is further compromised by the removal of certain sources that cities once depended upon. Checkpoints once allowed municipalities to tax goods entering their cities, but the removal of checkpoints has eliminated an important revenue source. Finding alternate funding sources has become increasingly important. Officials agree that municipalities should have the right to borrow money for funding, though they are currently denied that right. Other municipal problems that compromise urban revenue include corruption, inept tax collection methods, and pervasive tax fraud. Municipal funds are commonly directed into employee pockets instead of the services for which they are intended. Incompetent auditors have failed to detect financial improprieties in a timely fashion. Inadequate expenditure ceilings and improper accounting reports allow corruption to go undetected as well. Because the public often feels that corruption within the government prevents them from getting the services they deserve through their municipalities, people show unwillingness to pay taxes. Municipalities have not had a means to reprimand tax evaders effectively; consequently, tax evasion has become a significant threat to urban revenue. Persecution of tax evaders is often inconsistent due to delays and a lack of support from the courts. Officials have unanimously suggested that a municipal court be set up for the primary purpose of persecuting tax evaders. However, officials also note the need for clearer tax laws to aid in the efficiency and effectiveness of persecution. The overall opinion of officials is that urban finances need to be better regulated with clearer rules and regulations, financial goals, expenditure limits, and revenue quotas.

Residents agree that they need to pay for municipal services through taxes and that without taxes municipalities could not be held responsible for poor public service. They do complain, however, that many taxes are not in the state proclamations and thus cities lack the proper legislation to require tax payment. Residents also feel that taxes target certain segments of the population. Businesses operating legally shoulder the burden of taxes, while businesses without a license can avoid paying taxes completely. Residents note that municipal governments have not properly managed their financial contributions;

nor have residents had a say in how the governments use their contributions. Residents agree that the single most important improvement would be to allow the public to control municipal spending. They also desire a more systematic tax system that did not play favorites with individuals.

*Urban planning* Urban municipalities do not have a detailed and well thought out master plan for their city that systematically guides current land uses and future developments. This is due largely to the fact that urban plans are prepared and revised by a central agency, not by the cities themselves. Where some kind of a city plan exists, implementation of projects deviates from the initial plans to meet the personal needs of individuals. City administrators allow developments, in return for a kickback, that supersede the plan itself. The role of the Plan and Construction Department within cities is ill defined and poorly managed. Construction plans often lacked the detail and foresight needed to carry projects to completion. Zoning restrictions, building types, and standards were rarely considered. Like most sectors, their job had been simplified to concerns over land holding and management. The chain of bureaucracy from the *Wereda* to the Bureau of Works and Urban Development causes lease contracts to take enormous amounts of time, frustrating investors and discouraging investment within cities. Like other aspects of municipal government, the contracting process often lacks structure and proper guidelines. Moreover, the urban land lease system has opened up doors to fraud and abuses.

Residents also feel that decision-making processes regarding city planning have never involved them. Though they lack professional experience, they feel that they have much to offer in the form of a project vision that could meet the increasing needs of the community. They complain that current urban planning has failed to leave room for emergency problems or future needs, causing municipalities to be overwhelmed when problems arise. Planning has often been in response to investment trends and corporate pressures instead of to the public's needs. Instead of renovating historically diverse buildings, there is often a push to tear down and rebuild, threatening the historic and cultural aspects of the cities.

*Intergovernmental relations* The state and its various agencies as well as zonal and district administrations control cities through directives, rules and regulations. The accountability of cities to a chain of distant government structures has resulted in the lack of proper checking of municipal operation and management and has 'given room for corruption' (MWUD, 1999: 104) and indolence. City residents find the relationships between the various levels of government so bewildering that they cannot tell 'what powers resides in which authority' and what powers their city possesses, if at all, 'in relation to the various tiers of government' (MWUD, 1999: 105). The bureaucratic procedures of the cities are so complicated that they have provided plenty of opportunities for corruption.

Most residents stated that they did not understand the roles of various levels of government in municipal functions. They also did not understand the powers and functions of different sectors. Residents believe that municipalities should have authority over all their duties and should not be under the supervision of higher levels of government. Residents feel that higher levels of government should provide financial support and other resources to municipalities when needed, but should not attempt to manage municipalities.

*Service delivery* Cities are unable to cope with the rapid pace of their own growth and their environments are deteriorating. A host of problems plagues cities. Shortage of housing is a chronic problem as cities are unable to meet the demand for adequate quality and quantity of housing. Basic services are grossly inadequate in virtually all the cities. Sanitation facilities are inadequate and terribly poor. 'Open sewers, out-houses, and undisciplined solid waste disposal' are widespread. 'Human and animal defecations were almost in each corner poisoning the air with an unbearable stench and infectious droplets' (MWUD, 1999: 128). Most cities suffer from a shortage of water and have no liquid waste management to mention. Their roads are poorly constructed, lacking aqueducts, service lines, and sidewalks. Animals and pedestrians crowd the unclean streets. Other infrastructures such as electricity, telephones, and other public utilities are quite underdeveloped. Public health has reached crisis proportions. Corruption hampers the inadequate healthcare services. Malpractice and misdiagnoses abound. Medicines are in short supply and are too expensive and expired when available (MWUD, 1999: 128). These living conditions are closely associated with ill health and high incidence of diarrheal diseases, respiratory diseases, vector-borne diseases and other health problems connected with contaminated water supplies, inadequate sanitation, poor housing, and overcrowded conditions.

Most existing city infrastructures are deteriorating. Lack of maintenance causes city infrastructures to fall apart. Traffic signs, electrical lines, and telephone poles deteriorate. Sidewalks erode. Broken drainage pipes remain unfixed. Dirt plugs the drainage pipes and trash fills open ditches. New roads already need repair. Water pipes leak. Parks, where people dump their trash and defecate, houses thieves, stray dogs, and the homeless.

Because of the poor performance of the economy, poverty is worsening in all cities, bringing perturbing signs of social problems such as youth unrest, hooliganism, juvenile delinquency, and homelessness. Unfortunately, municipal response to these and other social problems is almost absent. Residents feel that the problem is not just a financial one. Poor management, inefficiency, incompetence, abuse, and misuse of financial resources and responsibilities are cited as major problems.

*Community relations* For the most part, community-city hall relations are not too good. City officials complain about residents' lack of serious involvement and participation in city affairs. Residents, on the other hand, say that their 'participation' is sought after decisions have already been made. For instance, budgets are passed without residents' comments and commitments. Issues affecting the cities are rarely presented to the public. City councils are more loyal to the ruling party line than to the people who elected them. Very few candidates for city government run on issues in the first place. The attitude of the dominant party shapes community forums and determines the performance of the city council. Most residents believe that political ideology corrupts any public discussion of city governance and management. Civil society organizations do not exist to pose challenges to the dominant party. In short, residents expressed a sense of voicelessness and powerlessness.

*Vision, mission, and growth of cities* Municipal leaders confessed that they are so bogged down with every day tasks that they have no time to pursue a vision or plan for their city's future, although they would like their cities to develop quickly. Residents wish their cities had had long-term development goals that might have prevented their decline over the past decades. Most residents have a pessimistic view of the future of their cities, save perhaps the residents of Bahir Dar and Kombolcha who are pleased about the spur of investments in their city in recent years and look forward to seeing even more growth and development in the future.

## Conclusion

The rate of urban growth far exceeds the capacities of municipal governments to provide and maintain the necessary urban amenities and infrastructural development. Urban centers suffer from many problems that accompany their growth and development, namely overcrowding, unemployment, poverty, and a deterioration of the physical and environmental quality of life resulting from the ever-widening gap between demand and supply of essential urban amenities and infrastructure such as housing, health facilities, water supply, sanitation services and transportation. The lack of basic urban services is a primary source of environmental stress and contributes to the steady decline of the health and welfare of urban dwellers. No matter what happens in the economic arena, continued urbanization will clearly form the pattern for a long time to come. Urban areas are still the best option for many who hope to better their way of life. Poor management of urban development in Ethiopia is largely responsible for the prevailing unhealthy urban living environment. The problems discussed in this chapter will not be ameliorated without the democratization of municipal governance, the active participation of urban residents, the incorporation of local community initiatives, and the synchronization of the various public sector services. Urban governance 'that brings to the fore the role that civil society plays and expands the range of stakeholders to include private sector agencies, nongovernmental organizations, community-based organizations, and a variety of interest groups' is imperative (Wekwete, 1997: 529). As long as urban citizens continue to have no confidence in the people in charge of municipal governments, the deterioration of urban living conditions will continue.

## Notes

[1] According to the 2000 Ethiopian *Demographic and Health Survey*, the median age at first marriage for women ages 20-49 is 16.4 years and for men ages 25-59 is 23.3 (CSA (2001a), *Demographic and Health Survey 2000*, Addis Ababa: CSA, p. 77).

[2] Cited in Assefa Haile Mariam and Helmut Kloos (1993), 'Population,' in *The Ecology of Health and Disease in Ethiopia*, Helmut Kloos and Zein Ahmed Zein (eds.), Boulder: Westview Press, p. 63.

[3] Cited in Kathryn A. Bard (1994), 'Environmental History of Early Axum,' in *New Trends in Ethiopian Studies: Papers of the Twelfth International Conference of Ethiopian Studies, Volume 1: Humanities and Human Resources*, Harold Marcus (ed.), Lawrenceville, N.J.: The Red Sea Press, p. 4; For a detailed discussion of the early urbanization processes in Ethiopia, see Akalou Wolde Michael (1969), *Urban Development in Ethiopia in Time and Space*, PhD dissertation, University of California at Los Angeles; Ronald J. Horvath (1969), "The Wandering Capitals of Ethiopia," *Journal of African History*, 10, 2: 205-219.

[4] The landscape of Addis Ababa is generally sloping down from north to south and incised by several steep sided valleys of waterways.

[5] For a detailed discussion of the evolution of the eucalyptus tree in Ethiopia, see Ronald J. Horvath (1968), 'Addis Ababa's Eucalyptus Forest,' *Journal of Ethiopian Studies*, vol. 6: 13-19.

[6] The 1907 decree gave away much of the city's land to the members of the royal family, Menelik's war chiefs, and other members of the aristocracy. Hence, a survey of land ownership conducted in Addis Ababa half a century later indicated 'that 58 percent of the total land area was owned by 1,768 large proprietors each with more than 10,000 square meters, or an average of 71,000 square meters per owner, whereas 24,590 small proprietors owning less than 10,000 square meters had only 7.4 percent of the total, the average size of such plots being a mere 150 square meters. [Another] 12.7 percent of the land belonged to the government and foreign embassies or legations.... A further 12 percent belong to the church while the remainder 9.9 percent...was largely royal land.' (Richard Pankhurst (1962), 'The Foundation and Growth of Addis Ababa to 1935,' *Ethiopia Observer*, 6, 1, p. 52). By 1910, the city had spread over 30 square kilometers (Eshetu Assen 1987: 81) with a population estimated at 65,000 (Richard Pankhurst 1962: 53).

[7] The size of the population of Addis Ababa has often in the past been a matter of conjecture. Estimates were often based on no more than pure speculation and individual expert impressions. Since 1984, however, extensive sample surveys and partial censuses have provided figures that are closer to reality, but certainly not unquestionable.

[8] Over the last ten years, youth crimes have skyrocketed in the city. The fact that the city's police force is notoriously corrupt has exacerbated the situation. In a 2002 conference on the deteriorating conditions of the capital city, a top-level government official made the most deprecating remark about the police force of the city. He said: 'There are an estimated 3,000 hardcore criminals in the city. The government hired 4,000 police personnel to deal with them. Now we have a total of 7,000 criminals.'

[9] The unemployment problem is exacerbated by the government's new educational policy that forces a vast number of teenagers to terminate their education at the age of 15 or 16. Under the new education policy, students who fail to make the grade to enter a two-year preparatory school (upon taking a ministry exam at the end of their tenth grade) are geared towards vocational education. The biggest problem, however, is that the policy was implemented before sufficient facilities and trained human resources were put in place across the country. The few vocational schools that are available are located in Addis Ababa and a few other major cities. Even where such schools are available, not all students can gain admission because of their inability to meet the required average grade point. The result is that tens of thousands of young adults across the country are forced to simply squander their lives doing practically nothing or get involved in criminal activities.

[10] The World Bank broadly defines urban informal sector employment as 'employment in urban areas in units that produce goods or services on a small scale with the primary objective of generating employment and income for those concerned. These units typically operate at a low level of organization, with little or no division between labor and capital as factors of production. Labor relations are based on casual employment, kinship, or social relationships rather than contractual arrangements' (World Bank (2003), *World Development Indicators*, New York: Oxford University Press, p. 71).

[11] Nearly all rural households own simple housing that are built using local materials and family and friends' labor. The overwhelming proportion of such housing, however, is overcrowded, unventilated, and with no access to potable drinking water, sanitation facility, and electricity.

[12] Cited in Solomon Gabre (1994), 'Urban Land Issues and Policies in Ethiopia,' in *Land Tenure and Land Policy in Ethiopia After the Derg*, Proceedings of the Second Workshop of the Land Tenure Project, Dessalegn Rahmato (ed.), Addis Ababa, Institute of Development Research, p. 282.

[13] The Kebele rent structure is also unfairly spewed to favor high-income tenants. The rent framework does not allow for accurate relationships between size and condition of the houses.

[14] In Addis Ababa, food and clothing generally absorb from two-thirds to four-fifths of a poor household's income. That does not leave much to improve or upgrade their dwellings.

[15] Cited in Shimellis Tekle-Tsadik (1998), *Organizing the Housing Sector in the Principles of Free Market Economy*, Addis Ababa: Addis Ababa Chamber of Commerce, p. 3.

[16] In 1996, 52 percent of housing in all urban areas were owner occupied, but it declined to 47 percent in 1998. Ninety-five percent of the rural housing is owner occupied (EPA (2001), *National Review Report on the Implementation of Agenda 21*, A report prepared for the United Nations Conference on Environment and Development, Johannesburg, South Africa, June 2002, Addis Ababa, EPA, p. 28).

[17] However, rather than solving the urban land ownership issue, the Addis Ababa City Administration seeks to discourage illegal settlements by bulldozing them away. In August 2002, the city administration bulldozed illegally built homes of more than 10,000 people in Bole Bulbula, near the city's International Airport, on the grounds that the settlements posed potential danger for aviation security as well as obstruct the planned development of the city. However, many have questioned whether the city's actions violated the evictees' human rights. Though the city may have the legal right to remove the dwellings, it failed to properly notify and consult residence before doing so. Many residents felt that they had an implicit understanding with the city that their shelters were not illegal, given that some had existed for many years and had even obtained electricity, telephone, and water connections. While realizing they have no legal claim to the land they had inhabited, residents felt that at the very least they should have been given proper notice and time to relocate to alternative location.

[18] Cited in Meheret Ayenew (1999), p. 13. There are between 100,000 and 200,000 street children in Ethiopia, according to the Ministry of Labor and Social Affairs. The UN Children's Fund (UNICEF) estimates that there are between 50,000 to 60,000 street children in Addis Ababa alone. At least one-quarter of them are girls (UN Office for the Coordination of Human Affairs, 'Ethiopia: Hundreds of street kids reportedly dumped in a forest,' 23 July 2002; See also, for detail, http://www.irinnews.org/report.asp?ReportID=28961). The city also has tens of thousands of beggars, many of whom have migrated from the countryside to make a living in the capital. Many of the beggars are homeless.

[19] AIDS is also a major cause of homelessness among children. With over 3 million people succumbing to the disease, more than a million children across the county have lost their parents. In the absence of institutional support, many of these orphaned youngsters end up living on the streets.

[20] There have been significant improvements in upgrading and expanding the country's road network in the last few years as the result of the implementation of the government's 10-year (1997-2007) infrastructure facilities development program. In 2002, 'design works for 32 roads covering 7,305 kms long have been completed, and construction work of upgrading on 13 asphalted roads of 2,097 kms long [was] ongoing. Major maintenance work has been carried out on main roads coverings 1,714 km' (World Bank (2001a), *Ethiopia: Interim Poverty Reduction Strategy Paper 2000/2001-2001/2002-2003*, Report No. 21796-ET, 30 January, p. 23).

[21] Cited in McGranahan, Simon Lewin, and et. al. (1999), *Environmental Change and Human Health in Countries of Africa, the Caribbean and the Pacific*, Stockholm: The Stockholm Environment Institute, p. 94.

[22] Every day at least 5,000 donkeys bring firewood, grains, and animal feed into Addis Ababa using the five main road arteries of the city (Tadesse, et. al. (2000), p. 35).

[23] The right to housing is recognized in various international human rights declarations and treaties, including the Universal Declaration of Human Rights of 1948 (article XXV, item 1); the international pact for Economic, Social and Cultural Rights of 1966 (article 11); the International Agreement for the Elimination of all Forms of Racial Discrimination of 1965 (article V); the Agreement for the Elimination of all Forms of Discrimination Against Women of 1979 (article 14.2, item h); the Agreement on the Rights of Children of 1989 (article 21, item 1); and the Human Settlements Declaration of Vancouver of 1976 (section III (8)) (Cited in Nelson S. Junior (2002), 'The Right to Housing and the Prevention of Forced Eviction in Brazil,' in *Holding Their Ground: Secure Land for the Urban Poor in Developing Countries*, Alain Durland-Lasserve and Lauren Royston (eds.), London: Earthscan, pp. 138-150).

# Chapter 3

# Urban Water Supply

The United Nations considers the provision of a safe and reliable water supply a basic human right. In March 1977, the United Nations Water Conference designated the period 1980 to 1990 as the 'International Drinking Water Supply and Sanitation Decade' with the hope to create a world in which there would be 'safe drinking and sanitation for all' by the end of the decade. The declaration attracted more international attention to global water and sanitation problems and encouraged investment by governments of developing nations and by bilateral and multilateral donors and non-governmental organizations. To that end, UN agencies such as the World Health Organization (WHO) and the United Nations Children's Fund (UNICEF), non-governmental organizations (NGOs) and local governments spent more than $100 billion on water development and sanitation projects in many developing nations of the world (World Resources Institute, 1998: 21). Over two decades, more than one and a half billion people gained access to safe drinking water for the first time. The proportion of people who received an improved water supply increased from 79 percent in 1990 to 82 percent in 2000. Even though sanitation did not have the same emphasis as water supply, nearly a billion people gained acceptable means of sanitation. The proportion of the global population with access to sanitation facilities increased from 55 percent in 1990 to 60 percent in 2000 (WHO/UNICEF, 2000: 1; Platt, 1998: 59).

On the whole, the UN declaration has contributed to a modest gain in access to improved water supply and sanitation facilities. However, the UN fell short of its goals. At the beginning of 2000, nearly 20 percent of the world's population was without access to safe drinking water (6 percent in urban areas and 29 percent in rural areas). This is largely because all efforts could not keep up with population growth and that many of the new water systems have failed because of poor construction, insufficient community involvement, lack of sustained political commitment, inadequate funding for operation, and maintenance and other system inefficiencies. The situation concerning sanitation is even more disquieting. Forty percent of the world population lacked access to proper sanitation facilities in 2000 (14 percent in urban and 62 percent in rural areas). The majority of these people live in sub-Saharan Africa and Asia, where less than two-thirds of Africans and one-fifth of Asians have access to some form of improved water supply. About 60 percent of Africans and 48 percent of Asians had access to improved sanitation in 2000 (WHO/UNICEF, 2000: 8). Hence, in spite of concerted effort and publicity, the 1980 International Water Supply and

Sanitation Decade has made little headway towards its goal of 'clean water and adequate sanitation for all' by 2001, let alone by 1990.

Moreover, the efforts of at least twenty years to improve water supply and sanitation services have not lead to substantial improvement in the health status of people in developing nations. Water-borne pathogens still contribute to four-fifth of all diseases in developing countries. The World Health Organization estimates that disease caused by water pathogens kills 9.1 million annually worldwide (Moore, 1999: 9.25), mostly in poor nations of the world. There are an estimated 4 billion cases of diarrhea each year (Lopez, 1996). Diarrhea kills roughly 2.2 million every year, 90 percent of the victims are children under the age of five (WHO, 2001b: 6).

Most Ethiopians can only dream of the availability of affordable, clean water for drinking on a constant basis. Most of the population, urban and rural alike, do not have adequate and safe access to potable water supplies and endure the concomitant environmental burdens. The government defines adequate water supply as 20 liters/capita/day within a range of 0.5-1 kilometer from home (World Bank, 1995: 3). Ethiopia has one of the lowest domestic water supplies in Africa. Only about 24 percent of the country's 62 million people had access to safe drinking water supply in 1998 (CSA, 1999: 158). By comparison, 33 percent of Ugandans, 49 percent of Kenyans, and 52 percent of Tanzanians have access to safe water (MOWR, 1998a: 20). At 83.5 percent, urban water supply coverage is better. However, less than one-fifth of the nation's population live in urban areas, and this coverage includes those who have low levels of service. Approximately 14 percent of rural households had access to safe drinking water in 1999.

Ethiopian public water supply agencies cannot meet the ever-increasing demands for safe and accessible drinking water. Much of the problem stems from the lack of proper and sufficient infrastructure for urban water supply. Water systems suffer structural and financial deficits and most rely heavily on external funds. They are not cost effective. Most of them do not recover all recurrent and capital costs (Kjellen and McGranahan, 1997). Past emphasis has been placed on the building of costly water infrastructure at the expense of efficient operation and management of the existing water systems.

## Water Resources in Ethiopia

With a surface area of 1,087,816 square kilometers, Ethiopia is blessed with water resources. It has the second largest water resources in Africa after the Democratic Republic of Congo. Fresh water comprises Ethiopia's most abundant natural resource. Eight major rivers systems—Abbay, Awash, Baro-Akobo, Dawa, Omo, Genale, Wabi Shebele, and Tekeze—transect the country's extensive surface. There are eleven major lakes in the country: Abaya, Abijata, Alemaya, Ashange, Awasa, Chamo, Hayik, Langano, Shala, Tana, and Zuway. Other inland water bodies include nine saline lakes (Abbe, Afambo, Gamari, Gaargori, Afrera, Asela,

Beseka, Chew Bahir, and Turkana) and four crater lakes (Hora, Bishoftu Guda, Zequala, and Wonchi). An estimated 1.3 trillion cubic meters of annual rainfall replenish the country's water sources (Waterbury, 2002: 16). The combined surface water potential of the country is estimated at over 112 billion cubic meters annually (MOWR, 1998c: 43). Groundwater is also plentiful in all but the most arid regions of the country. But Ethiopia uses only a tiny fraction of its huge water supply for drinking and other domestic purposes and for irrigation, hydropower generation, and industrial processes.

### River Basins

The *Blue Nile (Abbay) Basin* is the second largest basin in Ethiopia and occupies a catchment area of 199,812 square kilometers (that is, one-fifth of the country's land area). Over one-quarter of the country's population live in the basin, cultivating rain-fed crops on more than one-third of the higher grounds of the catchment area. The annual basin runoff of 52.6 billion cubic meters represents 50 percent of the total annual runoff of all rivers in the country.[1] The mean annual rainfall in the basin amounts to approximately 1,420 millimeters. The southern and southwestern sections receive the most rainfall (2,000-2,400 millimeters) and rainfall gets progressively less in a northeasterly direction (1,600 mm in the central section and 800 to 1,200 mm in the northeast). Many of the tributaries of the Blue Nile can be developed to generate electric power and/or develop irrigation schemes. Little of the estimated 430,000 hectares of irrigable land in the river valleys is under cultivation. Of the potential 20 or so hydroelectric plants that could be constructed in the basin, Ethiopia has only one project to show, the Fincha hydroelectric plant (Jovanovich, 1985: 84). In all, less than 3 percent of the Blue Nile water is retained within Ethiopia's boundaries (MOWR, 1997c: 1.5-1.7).

The *Tekeze River Basin* adjoins the Eritrean border in the north and drains the northern section of the northwestern highlands. The river is a principal headwater of the Atbara River in Sudan, a major tributary of the Nile north of Khartoum. The basin covers about 68,750 square kilometers with an annual volume of 7.63 billion cubic meters. Mountains surround the upper reaches of the Tekeze with elevations exceeding 2,000 meters above sea level. The low areas of the basin (below 1,500 meters above sea level), including 5,000 square kilometers of flatlands, are found near the Sudanese border, much of it located in the Eritrean territory. The mean annual rainfall in the basin is about 830 millimeters. However, the basin experiences considerable variations in rainfall, from 1,600 millimeters in the Simien Mountains in the central portion of the catchment to 500 millimeters along the Rift Valley escarpment to 350 millimeters in the western lowlands bordering Sudan. Over 90 percent of the waters of the Tekeze originate within the Ethiopian border. The Ethiopian side of the basin contains more than 250,000 hectares of land that can be irrigated (MOWR, 1996).

The *Baro-Akobo Basin* adjoins the Sudanese border in southwest Ethiopia and comprises a surface area of 74,102 square kilometers with an estimated volume of water of 12 billion cubic meters. The basin enjoys the highest average annual rainfall (1,600 mm) of any river basin in the country and, thus, possesses great potential for agricultural and water supply developments in the region (WRDA, 1995: 9).

The *Rift Valley Lakes Basin* comprises a surface area of 52,379 square kilometers of the southern part of the Ethiopian Rift Valley. Eight lakes form the principal feature of the valley. Altitude determines the quantity of rainfall, varying from less than 750 mm in the valley floor to 1,250 or more along the top of the valley scarps. The total water resource of the Rift Valley Lakes basin is estimated at 5.60 billion cubic meters (MOWR, 2002a: 5).

The *Awash River Basin* drains the northern part of the Rift Valley in Ethiopia. It is the only major river in the country that does not cross an international boundary; it drains into Lake Abbe, a salt lake on the border between Djibouti and Ethiopia. The basin occupies a catchment area of 112,696 square kilometers, but nearly half of this constitutes arid lands with little or no surface run-off to contribute to the river. The annual volume of the river is estimated at 4.6 billion cubic meters of which 678 million cubic meters are diverted for irrigation (MOWR, 1997d: 21; EVDSA, 1992: 38). The Awash River is the most intensively used for irrigation, energy production, and drinking.

In the south, the *Omo-Gibe Basin* covers an area of 78,243 square kilometers with 17.96 billion cubic meters of water. The basin enjoys a high amount of precipitation in excess of 1,500 millimeters annually in much of the upper portion of the basin and 500 millimeters in the lower basin.

In the southeast, the *Wabi Shebele Basin* has a surface area of 200,214 square kilometers. Much of the basin is dry land. While the mean annual precipitation over the whole basin is about 450 millimeters, the northern and western portions of the basin from which the river and its tributaries originate receive over 1,000 millimeters of rainfall annually. The annual surface water of the basin is estimated at 3.15 billion cubic meters and is characterized by large inter-annual variations in discharge.

The *Genale-Dawa-Weyib Basin* of 171,042 square kilometers consists of three major catchment areas that make up the principal tributaries of the Juba River in Somalia. Much of the basin, however, covers semi-arid to arid lowlands. The basin has a mean annual rainfall of 550 millimeters, but elevated areas of the northern portion of the basin receive an average annual rainfall amounting to 1,200 millimeters. The annual surface water in the basin is estimated at 5.8 billion cubic meters.

The country also has substantial amounts of *groundwater* with a potential safe yield estimated at over 2.6 billion cubic meters per year (UNEP, 1992: 60; MOWR, 1998a: 11-17). However, currently groundwater is a poorly studied and managed resource. In arid regions, groundwater is a major source of water to meet domestic and livestock watering needs. A number of industries also depend on groundwater supplies. Many thermal springs and wells are used for recreation and therapeutic purposes. Direct infiltration from surface run-off and rainfall is the primary source of groundwater recharge in the humid and partially humid regions of the country. Recharge in the more arid lowlands is by flood run-off and subsurface horizontal flow originating in highland regions. In general, aquifers in the country have moderate to low productivity. In the highlands, despite the high precipitation, aquifers are limited because of geology. Rocks tend to be older volcanic rocks with lower fracturing and higher weathering; thus, they tend to make poor aquifers. In the lowlands the lower rainfall limits recharge into aquifers. The groundwater recharge is no more than 20 percent of the precipitation falling in the highland regions and less than 5 percent in the arid regions. The Rift valley is one location where highly productive aquifers occur. Young rocks (with extensive fracturing) and unconsolidated sediments together make good aquifers with inter-granular permeability. Some of the groundwater in the region is characterized as hot groundwater and can be developed to generate energy (MOWR, 1998a: 15).

However, the distribution of both surface water and groundwater in the country is uneven—due to climatic, hydrological, and geomorphologic factors. Some regions have more water than they need; others have little or none. Major river basins in the eastern half of the country contain only 11 percent of the country's water resources, but over one-half of the population. In contrast, the river basins in the western half contain 89 percent of the water resources, but less than half of the population of the country (World Bank, 1996: 77). In addition, water supply suffers seasonal fluctuations as most rainfall occurs in one three-month wet season that is often unreliable. A feature shared by all the rivers in the country is that their discharge is subject to huge fluctuations, seasonally as well as annually. Over the last half a century, there has been evidence of increasing vagaries in precipitation, although whether the changes form a permanent climatic shift connected with global warming is disputed. Evidence seems to point to reduced average precipitation punctuated by severe drought episodes every decade or so. In spite of all this, the country has plenty of water to meet all its development needs. Almost all water problems are due to weaknesses in the institutional structure rather than scarcities imposed by nature.

## Ethiopia's Dilemma: Accessing Its Own Waters

Few major rivers in the world are contained within a single national territory. Ethiopian rivers are among the more than 200 rivers in the world that are shared by two or more countries (Frey, 1993: 55). All major rivers that originate in Ethiopia (except the Awash River) flow across the nation's borders and contribute about 75 percent of their flow to neighboring countries. The most important international rivers are those that form part of the Nile river basin: the Abbay (Blue Nile), Tekeze (Atbara), and Baro-Akobo (Sobat). These three rivers combined contain nearly two-thirds of the water resources of the country. In the south, the Omo-Gibe discharges into Lake Turkana which is located almost entirely in Kenya. In the southeast, the Genale-Dawa and Wabi Shebele drain through Somalia into the Indian Ocean. The Nile River is the biggest beneficiary of Ethiopian waters. The basin includes two sub-basins: the White Nile and the Blue Nile basins. The White Nile basin includes nine riparian states (Rwanda, Burundi, Tanzania, Democratic Republic of Congo, Kenya, Uganda, Ethiopia, and Sudan) before it joins the Blue Nile in Khartoum and then flows into Egypt. The Blue Nile basin includes two riparian states (Ethiopia and Sudan) and originates almost exclusively in Ethiopia. The White Nile/Bahr el-Jebel contributes a comparatively small proportion of the total of the Nile below Atbara, 15 out of 94.5 billion cubic meters, or about 16 percent. The remainder comes from the Ethiopian Highlands via the Tekeze-Atbara (13 percent), the Blue Nile (57 percent), and the Baro-Akobo-Sobat (14 percent (Mageed, 1994: 15).

Most water utilized in the Nile basin countries is surface water. Despite this, there are few international treaties that allow for joint management and development of the Nile River. Few agreements do exist, but these primarily favor Egypt's rights on the river, and to some extent Sudan's. The agreements sideline or exclude other riparian countries. In general, blatant disparity exists in the rights awarded to the different Nile basin nations, and Ethiopia is a major loser.

Ethiopia's rights are complicated by the historical precedence that downstream countries have enjoyed over the river and by the prior appropriation of these waters by Egypt and Sudan. The 1929 Nile Agreement between independent Egypt and Britain (on behalf of its colony, Sudan) recognized Sudan's rights to the Nile (4 billion cubic meters of water), but affirmed the 'historical ' and 'natural' rights of Egypt on the Nile (with an annual share of 48 billion cubic meters of water). The agreement underlined the rights of Egypt to oversee flows in Sudan, to initiate unilateral projects, and to disallow undertaking of any projects against her interests (Allan, 1992: 174). The agreement failed to mention other riparians' rights to a certain share of the water.

The agreement signed in 1959 between Egypt and Sudan also gave the bulk of the Nile waters to Egypt, a much smaller volume to Sudan, and nothing at all to upstream countries including Ethiopia. Of the estimated mean annual flow of 84 billion cubic meters at Aswam, the agreement allocated 55.5 billion cubic meters to Egypt and 18.5 billion cubic meters to Sudan. Evaporation and seepage were estimated to account for the remaining 10 billion cubic meters of water. The agreement does not provide for water to other riparian countries; however, both signatories agreed to study possible demands by others (Whittington and McClelland, 1992). Ethiopia and other riparian countries have repudiated the agreement. International water laws prohibit the exercise of absolute sovereignty over an international watercourse by upper riparian states. Neither can lower riparian states claim absolute territorial integrity on the premise that the natural flow of an international watercourse should not be impeded in the upper riparian states. Instead, the laws urge riparian states to partake in equitable and reasonable utilization of an international watercourse (Maurizio, 1997:122).

Egypt already uses every drop of the 1959 allotment and more. Both Egypt and Sudan plan to withdraw more water over and above the quantity stipulated by the 1959 agreement. Egypt's projected additional water need is estimated at 10 billion cubic meters a year (Befekadu, 1999). Egypt has recently unveiled a major project to pump nearly 6 billion gallons of Nile water a day from Lake Nasser into a 150-mile long canal, extending northwest across the Western Desert to irrigate 200,000 hectares of new tracts of farmland (Jehl, 1997). In undertaking this project, Egypt has not consulted with other riparian states. Sudan, too, has its own plan to expand its irrigation farmland by an additional 1.5 million hectares. This would require over 12 billion cubic meters of water above and beyond its allotment by the 1959 agreement (Befekadu, 1999). Sudan is also planning to build three new hydroelectric dams across the Nile in the north and is seeking foreign private companies to build, finance, and operate them.

Without a doubt, the water demand of the riparian countries will increase in the future. The total population of the Nile basin countries (including Burundi, Democratic Republic of Congo, Egypt, Eritrea, Ethiopia, Kenya, Rwanda, Sudan, Tanzania, and Uganda) was 307 million in 1999. It is projected to reach 512 million in 2025 (Postel, 2000: 47). Development in the region depends on agriculture. Agriculture currently uses as much as 80 percent of the water of the region even though much of this water is wasted through irrigation inefficiencies. The need for more food for the growing population will undoubtedly increase the demand for more water for irrigation. Increased urbanization will certainly impose heavy demands for more water. The potential of climatically induced adverse changes in the future water supply of the basin is significant. Thus, the issue of the use, management and ownership of the Nile waters will become important and controversial (Conway and Hulme, 1996: 278).

Egypt justifies its lion's share on the basis of historical rights or the principle of prior appropriation. Egypt wants to remain the largest beneficiary of the Nile River without contributing anything to its discharge. Egypt has always been extremely suspicious of Ethiopia and Sudan's interests in the Nile River, for it believes that any changes made upstream will adversely affect its own interests. In the past it has used military expeditions, bilateral agreements, diplomatic pressure and threat of force to gain more control over the waters of the Nile and to undermine Ethiopian and Sudanese interests on the river (*Addis Tribune*, 1996). But in recent years upper riparian states have challenged Egypt's continued dominance over the river. Ethiopia in particular has launched integrated water development plan studies on the Nile waters that originate in its sovereign territory. As a major riparian state, Ethiopia feels that it has the right to enjoy, within its territory, a reasonable and equitable share of all the benefits of the Nile waters without significantly harming the share of its lower riparian neighbors. Ethiopia seems unwilling to wait much longer to secure Nile water use while

Egypt continues to develop its desert reclamation policy (Whittington and McClelland, 1992).

Ethiopia pushes for regional cooperation and upholds the concept of equitable entitlement of the Nile waters in principle. Currently, cooperation and communication in the Nile Basin is being advanced through a series of international conferences held once a year in each riparian nation. While such conferences on Nile cooperation may have been helpful (especially in defusing disputes over water rights before they turn violent), they have yet to realize a fair and equitable distribution of the Nile waters among member countries. The Nile Conference in Addis Ababa, in June of 2000, saw an important shift from confrontation to cooperation among riparian countries, especially among Egypt, Sudan, and Ethiopia. Some sort of convergence is emerging among these three countries towards future cooperation in the areas of hydro-electric power generation, power linkages, management of the basin's wet lands, river regulation and erosion control. Nevertheless, Egypt and Sudan have not relinquished their 1959 agreement in which they claim all the waters of the Nile.

The lack of cooperation among the riparian states will present problems in the future when water demands are higher. The states must negotiate a more just and equitable distribution of the river's water. As a major user, Egypt will have to accept reality and enter into agreements with upstream riparian countries in return for a secure supply of water. Egypt has nothing to lose and much to gain by accommodating the interests of its co-riparians. Egypt will continue to enjoy a substantial share of water even under new basin-wide agreements. Additionally, improvements in, and better use of, water management technologies are estimated to save Egypt as much as 50 percent of its current use (Smith and Al-Rawahy, 1990: 220).

While pushing for a fair and equitable distribution of the waters it shares with its neighbors, Ethiopia can at the same time engage itself in developing a series of small-scale irrigation, hydroelectric power, and water supply projects on tributaries of its major rivers. Ethiopia does not necessarily need acquiescence from its co-riparian neighbors, nor does it need to wait for an international agreement to use more reasonably the waters that flow inside its sovereign political jurisdiction. Ethiopia should not view water development in terms of building large-scale dams. The construction and operating costs of large-scale water projects are exorbitant. Moreover, the projects invariably destroy the environment and ruin people's livelihoods. Large dams displace people, disrupt cultural practices, and wreck wildlife habitats, fertile farmlands, and grazing lands. Besides, any major water development will certainly require international financial borrowing. Ethiopia should know that such funding might not be forthcoming without basin cooperation. The country should instead go for small-scale dams as viable alternatives for supplying water and power to rural and urban communities. Small dams are by far more sustainable, cheaper, and have lower social and environmental impacts than large dams, and can be built with simple equipment and local labor and know-how. Small dams also hold great potential for the development of water-conserving irrigation schemes. Experiences have shown that many large-scale dam-based irrigation schemes have proven inefficient and environmentally detrimental: vast amounts of fertile farmlands and grazing lands have been destroyed because of waterlogging and salinization. Hence, the best option for Ethiopia is to develop small-scale water projects that are financially feasible, environmentally compatible, and socially acceptable.

Thanks to the cascading nature of many of its rivers, Ethiopia has rich hydroelectric power sources. The hydropower generation potential of its surface waters is estimated at 650 terawatt-hours per year (MOWR, 2002a: 5). Despite this, however, electricity is available only to approximately 11 percent of the households in the country. Only about 1.2 percent of rural households have access to electricity. Power is relatively more widely available in urban areas; 70 percent of urban households have it (CSA, 1999: 160). Over 95 percent of the households in urban areas of 50,000 and above have access to power for lighting, cooking and other household amenities. About 79 percent of the households in urban areas with populations between 20,000 and 49,999 and 57 percent of the households in urban areas with populations between 5,000 and 19,999 have access to power. For small towns with populations between 2,000 and 4,999, power reaches only 26 percent of the households. Regardless of the size of urban areas, however, power supplies are inadequate and intermittent.

## Water Resource Development and Management Policies

Up until 1992, the National Water Resources Commission (NWRC) managed the water resource development activities of the country. The NWRC was established in 1981 and had three authorities and one agency under its supervision. These were the Water Resources Development Authority (WRDA), the Water Supply and Sewerage Authority (WSSA), The Ethiopian Water Works Construction Authority (EWWCA), and the National Meteorological Service Agency (NMSA). There was also the Water Resources Council composed of high-ranking representatives of government ministries involved in water and related activities.

The establishment of the Federal Democratic Republic of Ethiopia in 1995 led to a major reorganization of the water resource management. The WSSA was decentralized and the regional bodies were given more administrative power. The EWWCA became a commercial enterprise and is now known as the Ethiopian Water Works Construction Enterprise (EWWCE). The Water Resources Commission was abolished. The former ministries, Water Resource Development Authority (WRDA), WSSA and Ethiopian Valleys Development Studies Authority (EVDSA), had most of their functions transferred to a new ministry, the Ministry of Water Resources (MOWR), replacing the Ministry of Natural Resources Development and Environmental Protection (MNRDEP) in 1995.

The country's constitution outlines provisions for the management of water resources. Water is a national asset; the state and the people of Ethiopia, as stipulated in Article 40(3), own all natural resources and land. Article 51(5) outlines the federal government's right to enact laws for the utilization and protection of natural resources. Article 52(d) outlines the rights of regional governments to administer their natural resources in accordance with federal laws (World Bank, 1996: 80).

At the federal level, the MOWR has the power to protect and manage the water resources of the country, formulate policies for long term planning strategies, set standards, and coordinate projects and their local or foreign funding sources. It also determines the methods and conditions for optimum allocation and utilization of water, issues permits to operate water works, determines water tariff and bulk charges for large-scale water supply schemes, prepares draft laws, and plans to protect and to utilize water resources for development purposes. The ministry is divided into several departments. The *Planning and Project Department* formulates the water resource policy and planning of the country and organizes other related government institutions and agencies around the policy and planning. It conducts a study of projects covered by the master plan studies and presents the plans to the federal government, regional governments, and investors. The *Basins Development Studies Department* conducts feasibility studies of the country's water resources, natural resources, land and socio-economic characteristics, and formulates water basin master plans and possible projects for the basins. The *Trans-boundary Rivers Study Department* conducts studies on trans-boundary rivers and proposes ways to utilize such rivers within the framework of the international water use principles. The *Design Department* conducts surveys and prepares designs for the federal government water works. It also provides regional governments technical and professional advice on water supply and irrigation work. The *Contract Administration Department* builds and puts to use studies of water works and design standards. It gives approval for professional accuracy of water resource development studies and design. The *Water Right Administration and Water Utilization Control Department* formulates and manages rules on appropriate distribution of water resources to regions and oversees the implementation of rules and works on improving them. The *Water Supply and Sewerage Department* prepares rules for the expansion of the water supply and sewerage services in the country. The *National Meteorological Service Agency* provides detailed information on the climate, weather, and atmospheric conditions for all the regions of the country.

At the regional level, though the federal government still maintains the right to promulgate laws, policies, and regulations on water supply and sanitation, the local water bureaus are the primary caretakers of the water supply and sanitation sector. Their functions include planning, designing and implementing water supply schemes, granting permits to agencies involved in water work construction, supervising the balanced distribution and utilization of water supplies, providing technical and maintenance support for water schemes in their jurisdictions, promoting community involvement, and ensuring that the directives of the federal government, specifically those aimed at conservation and utilization, are followed. They design schemes for rural water supply development, supervise and inspect water service units, provide technical and repair services to all water service points, and promote community participation. The regional government water bureaus also have a say in levying water tariffs and services fees and in formulating their investment plans within the budget outlines provided by the federal government (AAWSA/UNCHS, 2000: 13). Although decentralization is the best strategy for establishing better water supply schemes, it has not been coupled with improvements in the human and institutional capacities of the regional agencies that are now in charge of the sector. The federal government still remains responsible for many functions that the regional governments lack the capacity to carry out. These include, among others, policy setting, sector planning, financing, monitoring and enforcing environmental regulations, and providing training and developing research.

Other government institutions are also involved in water and sanitation related activities. The *Ministry of Health* (MOH) formulates public health policies at the national level and provides health and hygiene education to the populace. The *Ministry of Public Works and Urban Development and Housing* (MPWUDH) supervises government construction works and implements policies for urban development and housing. The *Environmental Protection Authority* (EPA) is a national federal body that formulates all policies geared towards managing and protecting the built and natural environment. It devises environmental protection standards and policies along with laws for their implementation.

In 1995, the federal government instituted a national policy to deal with the problems of water supply and sanitation in the country. Major aspects of the policy included promoting cost recovery and financing approach to help cover investment, operation and maintenance expenses of water supply; encouraging private sector involvement in such services as expert services (preparation of design, tenders, and feasibility studies), contracting (water well drilling, construction, and rehabilitation schemes), and sanitation service provision (solid waste disposal and sewerage services); providing performance incentives (bonuses, better payment, long leave, training opportunities, and promotion) to boost labor moral and efficiency of services; building regional capacity via organizational and technical management training to improve the efficiency and sustainability of services; and promoting stockholder participation in the maintenance and operation of water supply schemes (MOWR, 1995: 7-13). Its plan of action for 1993-1999 set goals for adequate water supply coverage of 35 percent for rural and 95 percent for urban areas; and 12 percent for rural and 70 percent for urban sanitation. The government defined adequate water supply coverage as 20 liters per capita per day within 0.5 to 1 kilometer distance from the source, even though the World Health Organization's standard are 35-40 liters/capita/day in urban areas and 15-25 liters/capita/day in rural areas.

However, progress in all these undertakings has been slow. Even though guidelines were put out to hike water tariffs commensurate to the costs of operation and maintenance, most urban areas have yet to implement new water tariffs. Private sector involvement and stockholder participation remain low. Management improvements were inconsequential. Despite some efforts in rehabilitating and improving the infrastructure associated with water supply and sanitation services in some urban regions, overall country coverage remains as grim as before, especially for sanitation services.

## Water Supply in Addis Ababa

Addis Ababa is encircled by the Entoto Hills in the north, Mount Menagesha in the west, and the volcanic cones of Mount Wachacha, Repi, and Furi in the southwest and west, respectively. For several years after its foundation, the city of Addis Ababa was supplied with water from a number of springs located at the foot of the Entoto Hills and hand-dug wells found at lower elevations of the city.[2] There are accounts of piped water being delivered to the city fifteen years after its establishment. A chronicle of Emperor Menelik II dutifully records the set up of piped water to the city center. This system was constructed as a cement canal and carried water all the way from the Entoto highlands. A big lime, cement, and coal reservoir, located in a field in Addis Ababa, stored water that was then pumped to pipelines that fed the city and palace compound.

The Qabbana was the first river used as a source of water supply for the city. Its flow was impounded by a structure of stones and *chika* that was located on the road to Hamre Noh Kidanemehret Church (at the crossing to the Ras Kassa residence). Ditches were dug along the right side of the river and carried water past the American Embassy and the Teferi Mekonen School to an open space by the entrance of the Crown Prince Assefa Wossen's residence. Here a purification and sedimentation structure with four walls was erected. The first wall allowed for sedimentation, the next two walls allowed for further purification, and the fourth wall was fitted with a wire gauge that served as a filter. The water was then distributed through five 2-inch pipes to various places in the city. These places included the residence of His Holiness Abune Matewos and the surrounding area, the Grand Palace and its surrounding area, the residence of Dejazmach Meshehsha Tilahun and the surrounding area, and the residence of Woizero Zewditu (later Empress Zewditu) and the surrounding area. Standpipes were erected to provide water to the residents living in proximity to the residences of the nobility.

The Arada district, the focus of all business activity, had another source of water. The river that flowed across the lands of Grazmach Dilnesahu and Bitweded Hailegiorgis in Qachane was dammed and its water was then purified and piped to the St. George Church, the residence of Bitweded Hailegiorgis and other people of the area.

Addis Ababa did not suffer water shortages at the time. There were numerous springs and ponds that residents cleared and fenced as the need arose. Deep wells were not necessary. The water supply system enjoyed improvements in 1910 when the spring at the foot of the road to Entoto Raguel Church was fitted with purification devices and its waters distributed to residence of the Crown Prince and to other neighboring areas. The Qabbana River, six springs, and two wells provided most of the water consumed by city residents.

In 1919, however, a water shortage occurred due to the lack of adequate rainfall. People began to dig wells along the river in the hope of tapping groundwater. This problem was acute until 1920 when torrential rains eased the situation. During the time of drought the government expanded its water supply scheme. Waters from the river that flowed across Ras Makonen's land were fitted with collection and purification devices and piped to the residences of many notable people in Qachane district, Arada Giorgis, Etegue Hotel district and Saratagna Safar. This new line was also a re-enforcement to the old line in Qachane.

In 1923 there were further improvements in the system as the Crown Prince and Regent (Later Emperor Haile Silassie) had advocated expansion and brought pipes when he visited Europe. The expanded line, built under the supervision of Ras Desta Damtew, carried water from the Entoto foothills to the Genete Leul Palace and surrounding areas, the residence of the Bishop, the Haile Silassie I Hospital, the Menelik II Hospital and the Empress Zewditu Hospital. The water quality of this line was excellent.

The Franco-Ethiopian Rail Road Company had also implemented a water supply system that was collected in Wachacha and distributed to the Qaranyo Medhanealem Church, the residences of His Holiness Abune Sawiros, Afe Negus Ketema, and other residents of the Sanga Tara District. All embassies, consulates, and legations also had access to piped and purified water. In 1931 the new Radio Station at Nifas Selk was also provided water supplies from the expanded system.

Since the time of their establishment water supply services were not charged; it was only sometime between 1926 and 1927 that water supply services were charged a fixed fee. This charge was levied at the time Ras Desta built a dam on the spring at the base of the mountain at Dil Ber and laid down a line to Arada. Further expansion of the system at this time also included two new reservoirs (each with 100 cubic meter capacity) and distribution of water to Gulale and Ras Mengesha districts.

In 1938, the Italians installed a water treatment plant at Shiro Meda to treat waters extracted from near by springs as well as from the Qabbana and Qachane rivers. A year later, they built a dam and a reservoir on the Little Akaki River at Gafarssa, to meet the increasing water demand of the city, with a potential of 10,000 cubic meters of water per day (Pankhurst, 1987: 131). In 1941, the total amount of water distributed for home consumption reached 2,800 cubic meters. Of this amount, 1,000 cubic meters were distributed untreated to the consumers. Increasing demand in the following years resulted in a greater expansion of the sector. After Ethiopia attained independence, some of the sector's functions were placed in the hands of the Municipality of Addis Ababa. The Municipality was reorganized and a water supply department was created under it.

At the time of Italian invasion a dam had been started on the Gafarssa Stream. It was never completed, but the facility could impound up to 1,500,000 cubic meters of water. The source remained unused, as there were no treatment facilities or pipelines to carry the water to the city. However, in 1950, pipelines were established to carry water to the city during the dry months when the water was relatively potable. The lines served the residents of Kofele district and a reservoir near the St. George Cathedral that provided water to other parts of the city. The establishment of the Gafarssa line alleviated water shortages in the lower parts of the city and could supply 3,000 cubic meters per day (though the pipelines allowed for only 2,000 cubic meters of flow per day).

The primary shortfalls of the overall sector were related to the fact that in districts where mains had high transmission capacity and where ample service connections existed, the sources were limited. By contrast, districts with adequate sources lacked the necessary infrastructure. In the meantime demand for water supply continued to grow and an additional 3,500 cubic meters of water were needed for the next 2-3 years. Existing sources and sector services were not sufficient. All the springs in upper Addis Ababa were already in use. In addition, these springs had depleting reserves due to the vast eucalyptus trees that were being planted. Neither did the wells prove to be as lucrative as had been hoped for. Of the 35 wells dug only 17 yielded water (at most 1,000 cubic meters in all). Thus, plans to direct the Gafarssa stream to the city were established.

In the mid-1950s, by increasing the height of the dam by seven meters, the capacity of the existing dam on the Gafarssa stream was expanded (by 7,500 cubic meters). A treatment plant that could treat 15,000 cubic meters of water per day was also proposed, and a 400-mm pipe would then carry this water to the city. The project was begun in 1953 and completed and inaugurated in 1955. With the advent of its use, older sources such as Kofele and Qabbana rivers were abandoned. Unfortunately, after a few years the dam was unable to cope with increasing city demands. Thus, the capacity of the treatment plant was doubled and the transmission lines were improved.

Addis Ababa's continuous demand for more water was a persistent problem. To counter this, the Lagadadi Dam Project and a possible sewage system were proposed in 1959. Three years of deliberation resulted in abandonment of the sewage system and a decision to focus on the dam project. However, studies conducted along with the American Technical Aid Program on the possible construction of this dam showed that there were many hurdles to overcome. A high power pump would have to be an essential part of the system so as to carry water from the dam site up to the city. In addition, the project would take five years to complete and estimates predicted that the city needed more water within three years. Thus, an alternative to the Lagadadi dam project was suggested.

A new dam, 4 kilometers from the old one, was to be constructed on the Gafarssa Stream and completed in three years. Then West German aid helped with preliminary studies that concluded in 1963. Once again this study did not come up with favorable results. There were many legitimate arguments against the project. First, an earthquake would destroy both dams. Second, both dams depended on the same watershed, making it likely that they would suffer from rainfall shortages together. Third, economic arguments showed that the project would cost about as much as the Lagadadi project; however, the dam itself would have only half the capacity. The Municipality was expected to cover local costs, but was financially unable to do so. Thus, the project was abandoned for the time being.

The dry season of 1964 was an especially difficult time. Water levels in the Gafarssa dam fell below the new spillway. The old spillway was used and three pumps (total capacity 13,000 cubic meters) were installed to pump water to the city. The amount of water being distributed dropped from 26,000 cubic meters to 10,000 cubic meters per day. The problem lasted 50 days and was remedied by the rainfall of the wet season. However, during the time it persisted, the Municipality launched programs for water conservation and the citizens of Addis Ababa met this with considerable cooperation.

The Lagadadi dam project was approved and its construction was to take approximately four years. During this time, temporary proposals to meet the growing demand of the city were suggested: using the river that flows from the Wachacha mountains to Sabata as an additional source (10,000 cubic meters of water per day), installing a sedimentation tank in the treatment facility and saving water that was used to wash and clean the facility (1,000 cubic meters), encouraging industries to dig wells for their water needs (300 cubic meters), and using the spring in Gafarssa valley as an additional source. Only the last two proposals were implemented.

A new supplementary reservoir built in the upper catchment of the Gafarssa Stream would supply the additional 800,000 cubic meters of water that were needed by the city in 1966. This project was to be 90 percent completed before the rainy season. As the Municipality was in need of other water sources and felt it could generate income from the completed project, it commenced work on the project. Unfortunately, the project was only 60 percent complete at the onset of the rainy season and though it did alleviate some of the water shortage (supplying 300 cubic meters) it did not meet its maximum capacity. The project was finally completed and inaugurated in 1966.

The completion of the largest dam, the Lagadadi, with a height of 44 meters and a capacity of nearly 44 million cubic meters, and a treatment plant in 1974 brought an additional supply of 50,000 cubic meters of water per day. An upgrade of the Lagadadi Treatment Plant in 1986 raised the output to 150,000 cubic meters per day. In 1997, the Gafarssa and Lagadadi treatment plants combined yielded over 142,000 cubic meters per day (Region 14 Administration, 1997: 18). A near drying-up of the Lagadadi reservoir and a 40 percent shortage of water in 1992 led to the construction of the Dire dam in a catchment adjacent to the Lagadadi in 1998 with the capacity of impounding estimated at 19 million cubic meters and a production of 40,000 cubic meters per day (AAWSA/UNCHS, 2000: 13).

Two treatment plants, located at Lagadadi and Gafarssa, process the city's water. The treatment processes include several steps. *Primary settling* involves hauling water into a large basin to allow particulate matter to settle. *Aeration* involves the agitation of the clarified water to promote the oxidation of easily oxidizable substances in water that would otherwise consume the chlorine or other disinfecting material to be added at the final stage of the treatment. *Coagulation* involves the removal of the finest particles, such as colloidal minerals, bacteria, pollen, and spores to give the water a clear appearance. *Flocculation* involves slow mixing of the coagulated water to encourage the formation of the floc masses and the entrapment of the suspended matters in the water to promote sedimentation and clarification. *Filtration* involves the removal of suspended material from water as it passes through beds of porous material. *Disinfection,* or *chlorination,* makes the water safe to drink by killing any bacteria and viruses that have escaped filtration and by preventing contamination during distribution system. Finally, the pH is adjusted (Bunce, 1994: 201-202; Koren, 1991: 216-218). The Gafarssa treatment plant processes 25,000 cubic meters per day while the Lagadadi processes 150,000 cubic meters per day. The two combined provide 88 percent of the city's total supply.

There are about 43 wells and 6 springs scattered in the city tapped to the supply system providing about 10,000 cubic meters of water per day, or 5 percent of the overall supply. The city also extracts groundwater from eleven wells in Akaki, 20 kilometers southeast of the city center, that yield a total of 13,000 cubic meters per day or 6.6 percent of the current supply volume. The location of these groundwater sources is highly vulnerable to pollution hazards emanating from the presence of several industries in the area. Many residents of the city, especially those in the outlying areas, use untreated spring water whereas wells serve as a source of water for industrial as well as for domestic uses. There are approximately 300 privately owned water wells at various locations in and around the peripheries of the city (Region 14 Administration, 1997: 98). The quantity of water supplied by these wells is unknown but the quality is considered quite poor. A survey of 45 wells in Addis Ababa showed that 19 were polluted with nitrates as high as 75 parts per million (ppm), well above the 45 ppm limit for drinking water (MOWR, 2001a: 97).

Established as an autonomous public agency in 1971, the Addis Ababa Water and Sewerage Authority (AAWSA) oversees the supply of potable water, waste water collection, treatment, and disposal in the city. In addition, it controls and protects the city's groundwater, dams, catchment areas, and waterlines from which it draws water for the city from contamination (AAWSA, 1984: 4). Since all the sources of water for the city are located outside the jurisdiction of the city, AAWSA has to seek the permission of the Oromiya Regional State to draw water for the city. The Authority carries its tasks through sub-divisions. The *Water Service Department* monitors the treatment, storage, and distribution of drinking water in the primary lines. The *Central Laboratory Service* ensures the water meets the quality defined in drinking water standards. The *Sewerage Services Department* oversees liquid waste collection and treatment and maintains sewer lines and private sewer line connections. The *Engineering Services Department* oversees water distribution networks, water meter reading, and inspection and vehicle maintenance (Johannson, 1996: 11).

At present, the demand for water in the city far exceeds the supply, perhaps by as much as 29 percent (Region 14 Administration, 1997: 98). To meet the current and future water needs of a fast growing population, which is projected to reach over 6 million in 2020, AAWSA has to look for additional water sources. The projected demand for potable water in the city in the year 2020 is over 500,000 cubic meters per day. The city plans to impound surface water from the Sibilu and Gerbi rivers located about 30 and 45 kilometers to the north of Addis Ababa, respectively, in Oromiya region. The combined yield from these two water projects is estimated at 700,000 cubic meters of treated water per day by the year 2011 (Adam and Mohammed, 1997: 2). Improvement on the Gafarssa dam is estimated to yield another 43,000 cubic meters per day by the year 2017. The city's plan also foresees an additional output of 60,000 cubic meters of water per day resulting from replacement of pipes and maintenance efforts (World Bank, 1995: 35).

## Water Supply Problems in Urban Areas

In 1998, only 23.7 percent of all households in Ethiopia had access to safe drinking water (10.2 percent from protected wells/springs, 10.8 percent from public taps, and 2.7 percent, from private taps) and the remaining had only water from wells and surface water sources (Table 3.1). In urban areas, about 73 percent accessed tap water, but most of this water (54 percent) came from public taps. Addis Ababa had the highest proportion of households with private or yard connections (38 percent) followed by Harari (City) State, 28.9 percent; SNNPR, 13.4 percent; Oromiya urban areas, 11.6 percent; Amhara urban areas, 10.4 percent, Dire Dawa, 11.3 percent; and Tigray, Afar and Somali urban areas with less than 10 percent. However, the majority of the population in major cities and towns use clean water for drinking and other household purposes. Overall, household access to tap water (private, yard, and public taps combined) varied widely among the regional states of the country. The federal capital, Addis Ababa, ranks first in accessing tap water (88 percent of the households), and the regional states of Afar, Somali, Gambela and Benishangul-Gumuz rank lowest, with 11 percent, 16 percent, 17 percent, and 12 percent, respectively (CSA, 1999:

170-172). Urban piped water service comes in three forms: private connections, yard connections, and public tapes or standpipes. Private connections exist inside the residence and may indicate the presence of sanitation facilities within the residence that make use of water. Yard connections deliver tap water to a residential compound. More often than not, neighbors share yard taps; often a single yard tap serves several households. About 38 percent of the households in Addis Ababa have either private or yard connections.

Public tapes commonly serve those households without private or yard connections. Such structures serve large proportions of the urban communities. In 1998, over 87 percent of the people in Dire Dawa (the second largest city in the country), 74 percent in Adama, 72 percent in Jijiga, 68 percent in Bahir Dar, 67 percent in Dessie, and 63 percent in Addis Ababa accessed safe drinking water from public fountains (Table 3.2). Over 11 percent of the population in these and other large cities and towns obtained water from rivers, lakes, or unprotected springs and wells (CSA, 1999: 172).

A small urban area will have less safe drinking water than a large urban area. Only 60 percent of the people in towns with a population of 2,000 to 5,000 enjoy safe water; whereas, 92 percent of the people in cities with a population of 50,000 or above enjoy safe water. The primary sources of drinking water for smaller towns are ground water and surface water. Groundwater comes from wells that tap into underground aquifers. Most wells are excavated by hand and

**Table 3.1  Sources of Drinking Water by Place of Residence, 1998**

| Sources of Drinking Water | Place of Residence/Households | | |
|---|---|---|---|
| | Country Level (%) | Rural (%) | Urban (%) |
| Protected Well/Springs | 10.2 | 10.1 | 10.6 |
| Public Tap | 10.8 | 3.6 | 54.1 |
| Own Taps | 2.7 | - | 18.8 |
| Rivers/Lakes | 43.5 | 49.5 | 7.0 |
| Unprotected | 28.2 | 32.2 | 4.1 |
| Others | 4.7 | 4.6 | 4.4 |

Source: Central Statistical Authority (1999), *Report on the 1998 Welfare Monitoring Survey*, Statistical Bulletin No. 224, Addis Ababa: CSA, p. 158.

**Table 3.2  Distribution of Households by Sources of Drinking Water in Selected Urban Centers, 1998**

| Selected Urban Centers | Sources of Drinking Water | | | | | |
|---|---|---|---|---|---|---|
| | River, Lake (%) | Protected Well /spring (%) | Unprotected Well / spring (%) | Public Tap (Bono) (%) | Own Tap (%) | Others (%) |
| Mekele | 1 | 8.7 | 2.7 | 58.9 | 28.8 | - |
| Assayita | 12.7 | 6.4 | 4.3 | 56 | 15.7 | 4.9 |
| Gondar | 10.8 | 9.1 | 3.4 | 52.2 | 23.5 | 1.2 |
| Bahir Dar | 3.2 | 0.8 | 2.3 | 68.3 | 25.5 | - |
| Dessie | 1 | 1.2 | 0.7 | 67.2 | 24.7 | 5.2 |
| Jimma | 1.4 | 16.2 | 5 | 56.7 | 11.1 | 9.5 |
| Adama | 0.6 | 0.9 | - | 74.2 | 23.2 | 1.1 |
| Bishoftu | - | 1.8 | 0.4 | 58.5 | 30.9 | 8.4 |
| Jijiga | - | 1.5 | 1.2 | 72.1 | 11.1 | 14 |
| Asosa | 15.8 | 32.8 | 23.2 | 19.4 | 6.2 | 2.6 |
| Awasa | 0.6 | 3.5 | - | 74.7 | 21.2 | - |
| Gambela | 33.9 | 5 | 30.7 | 47.8 | 6.3 | 3.3 |
| Harari | 0.4 | 4.3 | 2.3 | 63.3 | 28.9 | 0.8 |
| Addis Ababa | 1.2 | 1.1 | - | 50.7 | 37.6 | 9.4 |
| Dire Dawa | 0.4 | - | - | 87.4 | 11.3 | 0.8 |
| Amhara other Urban | 9.6 | 17.6 | 7.2 | 53 | 10.4 | 2.2 |
| Oromiya other Urban | 13.6 | 10.6 | 2.5 | 57.7 | 11.6 | 4 |
| SNNPR other Urban | 9.3 | 26.9 | 5.2 | 42.4 | 13.4 | 2.7 |
| Other Urban | 4.2 | 14.2 | 15.1 | 47 | 8.6 | 11 |

Source: Central Statistical Authority (1999), *Report of the 1998 Welfare Monitoring Survey*, Statistical Bulletin No. 224, Addis Ababa: CSA, p. 172.

those found in wet regions are usually shallow, extending to 2 to 6 meters beneath the surface of the ground deep enough to reach the water table. Others, especially those found in dry regions, are deep—extending beyond 6 meters into the ground—and are dug with the use of power equipment.[3] Springs provide another source of drinking water for many towns. Spring water is groundwater that outcrops where the underground aquifer traverses the surface of the earth. Sources of surface water include rivers, lakes, and reservoirs. During the rainy season, people depend less on the town water system or on water vendors for their household needs. Many urban households, especially the poor, collect and store rainwater in clay pots, plastic containers, or metal drums. Such water is used for all kinds of domestic purposes such as bathing, washing clothes and utensils, and housecleaning.

In many cities and towns the shortage of water outlets has given rise to a substantial private water market. The private water market consists of different kinds of vendors. There are vendors who buy water from municipal stand pipes or draw water from rivers or unprotected wells or springs and sell it to homes located in areas without piped service or with very irregular service. These vendors transport water in a steel drum with two rings attached around it to facilitate rolling on a flat surface. Some vendors haul water on pack animals, usually donkeys. Other vendors consist of households that have connections to the system and sell water to low-income neighboring consumers. In the city of Harar, for instance, an estimated 30 percent of the households with a house or yard connection sell water to their neighbors (MOWR/Ernst &Young, 1997a: 28). In addition, mobile vendors buy water from connected households and transport it to consumers. Thus, in poor urban communities that lack private connections and cannot access public fountains, people have no choice but to purchase water from private vendors at steep prices. Private vendors levy much higher charges on water (2.5 to 10 Birr per cubic meter) that they receive at the government tariff rate. Surveys of 295 urban areas across regional states indicated that half of the urban areas paid more than 10 Birr per cubic meter of vendor water. In some urban areas, a jerry can of water (20 liters) cost as much as 2 Birr, which translates to 100 Birr per cubic meter of water. The average vendor price for urban areas with 30,000 or more people is 4.13 Birr while it is 10.29 Birr for those urban areas with below 30,000 population (that is, in 279 of the 295 urban settlements surveyed). The average vendor tariffs also vary from one regional state to the other. At 21.58 Birr per cubic meter of water, the Somali State has the most expensive water price. Water is very scarce in the State due to extensive aridity. The Tigray State has the lowest average vendor tariff, at 2.67 Birr per cubic meter. Other States fall in between these extremes: the Afar and Amhara states average about 4.00 Birr and the Gambela, Harari, SNNPR, Benishangul-Gumuz and Oromiya states average between 9.00 Birr and 14.00 Birr (MOWR/Ernst and Young, 1997b: 23-25).

In most towns, interruptions of water supply constitute not the exception but the norm. Town water systems run sporadically: some run for only a few hours every day; others run every other day during day light hours. Upper class households can better cope with the situation than poor households; when the system runs, they usually fill storage tanks for use when the system is down. As a result, these households rarely suffer interruption of water supply for any length of time. Water is also rationed in towns where supply is short. In such situations, private connections and public taps operate in shift systems. Consumers generally do not know in advance how long water supplies will be terminated.

In nearly all cities and towns in the country, water sources and services cannot satisfy the water demand. Rapid population growth has exacerbated the discrepancy between water demand and supply and the current water demand outstrips the designed capacity of the water supply. Many towns distribute water to different sections of the town on alternating days. Because of poorly designed facilities and bad estimates on population growth, water demand rate, and the rate of recharge of sources, many water systems have exceeded their design life (MOWR, 1997e). As a consequence, providing adequate water supplies for some urban areas with larger population has become difficult and expensive. The old city of Harar, for instance, has already tapped all existing water supplies in its vicinity and must now search for water from great distances. In the good old days, much of this city's water came from Lake Alemaya, some 17 kilometers away. The lake is now fast vanishing from the face of the Earth—like nearby lake Adele before it—due largely to a rate of withdrawal that far exceeded the lake's natural capacity to recharge. Furthermore, intensive farming activities in the hills surrounding the lake have caused sedimentation. Throughout much of the year 2002, water was not supplied in the city throughout the day, but only during peak hours of early morning and early evenings. Often, the system delivered little or no water for five or more days in a row. Having ignored the looming water crisis for years, the Harari Regional government clambered (in 2002) to find financial resources to pipe water from a low altitude aquifer located at Dire Jara near the city of Dire Dawa 55 kilometers away. According to estimates, the Dire Jara well field will last only through 2012. Another plan to exploit an aquifer located 27 kilometers west of Dire Dawa is also being considered. It is believed that this aquifer will last through 2022 (MOWR, 2002a: 9).

Where the rate of developed facilities is insufficient to meet the needs of a rapidly growing population, wells or boreholes equipped with hand pumps are often used as an alternative to the piped water supply. Where such alternatives are not available water is obtained from unimproved water supplies (for example, rivers, ponds, lakes, and unprotected springs)—often from the nearest source irrespective of water quality. Young women and children shoulder the burden of supplying and transporting water from alternative sources to the family, regardless whether the path is long and the climb is steep. A 1997 household survey of eight towns—Alaba Qulito, Assayita, Butajira, Deghabur, Dilla, Qabri Dehar, Harar and Wolkite—showed that women collect 63.7 percent of water; young children, 29.6 percent; and adult men only 6.7 percent. Time invested in water fetching varies from town to town depending on sources and availability.

Where public fountain water is available in adequate quantities, time consumed for collection is relatively small. On the other hand, much more time is spent where water is scarce. For the eight towns, on the average, it took one-half to an hour per day for about 58 percent of the households and more than an hour for the remaining households (MOWR/Ernst & Young, 1997a: 5.4).

Although municipal water agencies and the federal government have attempted to improve the quality and coverage of water supply services, their improvements have not kept pace with the increasing water demand of urban areas. Poor management of the majority of municipal water systems constitutes the chief problem of the country's water sector. Many problems contribute to the failure of water supply systems in the country, including:

*Inadequate operation and maintenance*  For municipal water agencies, operation and maintenance expenditures include salary, the replacement of pipes, valves, water meters, and service vehicles. Whereas salaries are regularly paid, regular maintenance of the water systems is the exception rather than the norm. When external funding is forthcoming, municipalities opt to make investment in constructing new systems instead of rehabilitating existing systems. Most urban water systems have old equipment and pipelines that receive no preventive maintenance. In many towns public water fountains suffer frequent breakdowns because of too much use and abuse. However, repairs on public water taps in cities and towns are either done on an untimely fashion or not done at all. Boreholes and pumps used by many systems are not attended to unless they manifest drastically reduced output. For instance, of the 265 public taps that Gondar (a town of 156,000 population) had, only 35 were working in 1995 (AAWSA, 1993: 15). In Dilla (pop. 42,000), of the nine public fountains, five were not functioning. Of the 15 public fountains in Deghabur (pop. 34,000), only four were functioning. In Harar (pop. 101,000) all 8 public fountains were not working in 1997 (MOWR/Ernst & Young, 1997a: 28). The problem with most public fountains is that they lacked repairs, although in some cases, as in Deghabur, water vendors intentionally wreck taps to boost their income.

There are thus no systematic approaches to maintain and operate water systems. Most maintenance is done only after the system breaks down completely. A 1995 World Bank study showed that four-fifths of 200 urban schemes were in need of overdue rehabilitation had low production and high leakage problems. Thirty percent of the 6,000 rural water supply schemes were also in need of repair (World Bank, 1995). Maintenance suffered from insufficient funds to buy spare parts and to hire technically qualified personnel. Almost 70 percent of equipment used in the water sector is imported from abroad. The procurement of this equipment is time consuming, taking almost a year to acquire from the time of tender announcement to the date of receipt. Steps involved in this procedure include the tender process, analysis, selection, announcement, negotiation, contract signing, correspondence, transportation, port clearance, and tax payment (MOWR, 1998b: 45-50). Low staff morale exacerbates matters as it encourages poor performance. The unfortunate consequence of all this is that when water systems break, people go without water service for weeks if not months. In the mean time, they must use water of suspect quality and make do with far less than they require; thus, they expose themselves to health risks, including intestinal worms, diarrhea and other diseases that could be prevented through an adequate water supply.

*Unaccounted system losses and leakages* The amount of water losses in many piped water systems of towns is staggering. In most cities and towns an estimated 30-40 percent of the total quantity of water produced never makes it to the consumer at the end of the pipe because of leaks. In towns like Adwa, Awash, Debre Berhan, Bishoftu, Fitche, Metahara, Negele Arsi, and Sabata, unaccounted-for-water (that is, the difference between the quantity of water produced and the quantity of water billed) ranges between 40 and 60 percent (MOWR, 1998a: 21). Addis Ababa loses through leakage approximately 40 percent of the 198,000 cubic meters of water it produces. In all, nearly 29 million cubic meters of potable water is lost in the system every year, enough to meet the water needs of the city for about six months or more than enough to supply all of Dire Dawa, the country's second largest city, for the entire year. At a production cost of 3 Birr per cubic meter, the loss amounts to 87 million Birr or over US $10 million annually (AAWSA/UNCHS, 2000: 4). Leakage causes not only water loss but also water contamination.

Most of this water is lost through leaking pipes, unauthorized use, incorrect connections, incorrect meter calibration or reading, overflowing service reservoirs after abstraction, treatment, or during distribution, and inadequate system control. A 2000 study of the Addis Ababa water supply networks found that as much as 85 percent of the leakage 'occur at connections and 37 percent of the damages causing the leaks were due to improper fitting' (AAWSA/UNCHS, 2000: 47-48). Leaks also occur because connections to reservoirs in households and establishments are not equipped with automatic shut-off valves. Individual households install reservoirs because of irregular distribution caused by limited municipal storage capacity, heavy demand fluctuations, variations in rainfall, and high distribution losses.

*Inadequate cost recovery* Any public utility must fully recover its operational, maintenance, investment, and depreciation costs. However, urban and rural public water prices in Ethiopia are inordinately low; they do not reflect the true cost of water production. Over 80 percent of the major urban supply schemes have water tariffs of 1 Birr or less per cubic meter. Addis Ababa's water tariff, for instance, remained at 0.50 Birr per cubic meter for over four decades while the costs of operation and maintenance as well as capital investments appreciably augmented over time. Additionally, the tariff was uniformly applied to all consumers without taking into account the variations in households' ability to pay. It is no wonder that the city water supply authority remained financially insolvent throughout its existence.

In most towns the cost of metered water range between 0.50 Birr and 1 Birr per cubic meter whereas the cost of production range between 1.10 Birr and 2.50 Birr. However, a 1997 survey of 297 towns show that water tariffs varied by region, size of city or town population, and the type of service. The average price for privately connected water is 0.84 Birr per cubic meter for towns with more than 30,000 population while the average tariff is twice as much (1.62 Birr) for urban areas with less than 30,000 population. The findings suggest smaller urban areas with lower levels of water service pay considerably more than larger urban areas. The average tariff for privately connected water is the highest in urban areas in the Somali State (4.59 Birr), followed by Benishangul-Gumuz (2.50 Birr), Tigray (1.82 Birr), and the Afar, Oromiya, the SNNPR, and Amhara states, with tariffs ranging between 1.36 Birr and 1.48 Birr. The Harari State and Dire Dawa have the lowest tariffs, with 0.50 Birr and 0.88 Birr, respectively. The average tariff for public tap users in 311 urban areas ranges between 0.88 Birr in Dire Dawa and 3.00 Birr in Gambela. Urban areas with 30,000 or more people pay an average public tap tariff of 0.92 Birr while those with 300,000 or less pay 1.62 Birr on the average (MOWR/Ernst & Young, 1997b: 21-27).

In most cases, water tariffs do not cover the cost of maintenance and operation. Most towns and regions lack systematized cost data, even though experience in some areas indicates that tariffs of at least 2 to 3 Birr are needed for direct cost of recovery of electrical systems. These tariffs cover only maintenance and operation, not capital depreciation or future investment. Diesel based systems require tariffs between 2.75 to 4 Birr to cover direct costs. Operation and maintenance costs as reported by town and regional bureaus do not include material and service costs. Nor do they include staff wages, which are usually low and hence dampen staff morale and motivation. Transportation and recurrent and depreciation costs are also not factored into expenditures (MOWR, 1997e: 10-11).

Municipal authorities justify the prevailing low tariffs on the ground that the urban poor, which constitute the vast majority of the urban population, cannot afford to pay rates that are based on the real cost of water production. This justification, however, does not stand up to scrutiny; the fact remains that most urban poor households pay many times more per unit of water than rich households. Most urban poor have no pipe water connection to their homes or compounds; they purchase water from public fountains, their neighbors, or from water vendors who charge anywhere from 5 to 20 times the public rate. In Addis Ababa, for example, water from vendors costs 5 times the public rate; in Shashamene, Jimma, Dessie, it costs 7, 11, and 12 times higher, respectively. A World Bank survey in 1995 showed that most people in Ethiopia would pay more in tariff if they were guaranteed better water quality and easy access to it. The study found that many already paid secondary water supply vendors as much as 25 Birr (in the dry season) and 12.5 Birr (in the wet season) per cubic meter of water (World Bank, 1995: 41).

When tariffs are low they adversely affect the sector's financial and, hence, overall structure—there are no funds to expand the network and the poor are permanently excluded from service benefits. This situation leaves the poor with inadequate water supply and can further exacerbate matters because a constrained supply can lead to higher prices. As a result, the poor may ultimately pay more for water than the rich. Thus cost recovery via consumer payments may have several advantages. First, it would help the sector recover costs that could be used to expand and improve the system in poor areas. Second, public funds would less likely be used to service the water needs of well-to-do consumers. Third, water supply sectors would be more in tune with the genuine needs of the paying consumers. Fourth, far less water would be lost to leaks—currently a major problem in urban water systems throughout the country—as water agencies are able to maintain water networks. Finally, it would lessen the sector's dependency on public and external donor funds (Kjellen and McGranahan, 1997: 43; Cairncross and Feachem, 1993: 55).

The low water tariff encourages extravagant consumption and unnecessary wastage. Households waste water in direct proportion to their income. In low-income households, water is usually precious and so very little is wasted on unessential uses and the loss is predictably low. On the other hand, in high-income households, consumption of water is usually high and wastage on what are perceived to be unimportant uses is also likely to be high. Thus an increased water tariff could promote water conservation amongst the wealthier users. This can be done by introducing incremental block pricing systems where tariff rates increase in step per pre-determined volume of water usage. In other words, low tariffs are charged for the initial block of consumption and increasing tariffs for extra blocks. Many developing countries use block tariffs to subsidize poor households. Such a tariff system can have two benefits. First, they oblige large consumers such as firms and high-income households to subsidize the consumption by low-income households. Second, they discourage waste and promote conservation. Whereas low volume users will not necessarily change their demand habits, high volume users, such as industries and other commercial enterprises, are likely to reduce their consumption rates in response to higher prices. However, these advantages can only be realized if low-income households are served by direct waterlines. In most cases high connection costs prohibit low-income households from enjoying municipal safe drinking water. One method to alleviate high connection costs is to provide credit to low-income households. Subsidizing all or part of the connection charge might be another possible option (World Bank, 2000c: 84; MOWR, 1998d: 1-2).

In 1995, the federal government issued 'Basic Principles and Guidelines for Water Tariff Study' that underlined the need for tariff restructuring in order to recover some or all of the cost of running the water sector. Some regional governments (for example, the Amhara and Oromiya regional states) have implemented increased tariffs in a few urban areas. However, except in Addis Ababa, well-defined tariff structures have yet to be universally implemented across the country and, thus, the water sector remains financially unviable (World Bank, 1995: 7).

*Water wastage* Wastage plagues the water supply sector. Public institutions (government ministries and agencies, schools, and public healthcare institutions and hotels), the worst wasters of water, consume much water for which municipal water agencies are not paid. Leaks in bathrooms and cafeterias are rarely fixed in time to save water. Sometimes it takes flooding to attract the attention of the people in charge of repair. Oddly, few employees of public institutions care to report leaks. Imposing a water tariff on non-paying public institutions would discourage their abuse of precious water. Water wastage is also common in homes hooked up to municipal water mains because it is obtained so cheaply.

*Inadequate distribution network* The existence of a town water system does not necessarily guarantee access to water. Connection costs comprise an important part of water affordability. Only those who can afford the connection fees can use the service. Poor urban households have difficulties generating enough capital to pay connection fees. Initial connection fees amount to three to ten times the average household income in Butajira, Alaba Qulito, Dilla, Wolkite, Assayita, Harar, Deghabur, and Qabri Dehar. Subsidizing connections costs to low income households can make it easier for the urban poor to access safe drinking water. Studies link home water sources with improvements in household health (Kjellen and McGranahan, 1997).

*Poor water quality* The quality of urban water is as much a concern as quantity. Water intended for human consumption must be free from concentration of chemical substances and organisms that may be harmful to health. Urban areas produce huge volumes of sewage, from both households and industries. This poses immense problems for both human and environmental health. Water-related diseases are widely prevalent in urban areas largely owing to insufficient water and poor sanitation services.

      Generally, most of the efforts of municipal water agencies are devoted to producing as adequate water as possible for the urban population. As a result, the quality aspect of water is often given less attention than is necessary. There is no systematic and efficient surveillance system in place to test the quality of water at all locations, from the sources of raw water to the household stage. Water lines are attended to only when there is a major breakage. Regional bureaus are supposed to perform water quality (chemical and bacteriological) tests on an initial and request basis. However, periodic (annual/monthly) checks are not the norm. There are no attempts to test water quality at the town level (MOWR, 1997e). For many towns, water supplies involve treatment and pumping. Often these processes are vulnerable to failure. Diesel fuel is not readily available for many towns. Disinfectant (for instance, chlorine which can kill most waterborne microorganisms) may also be in short supply.

Over all, there are no regular approaches to monitoring and controlling water quality. The legal and institutional framework does not support any monitoring schemes. This is especially so with respect to industrial and domestic liquid waste discharging into rivers and streams. Water quality control is supposedly the responsibility of the agency that supplies it. At present neither local nor regional water agencies have the technical skills, proper equipment, or laboratories to control water quality. In the past, central governmental institutions were associated with the water sector: the Ministry of Water Resources, the Ministry of Health, and the Ethiopian Standardization Authority (ESA). The MOWR's duties included the management and prescription of quality standards for water use. The MOH was responsible for prevention and control of pollution of streams and rivers. The ESA issued a standard for water quality. However, the fragmentation of duties amongst the agencies proved ineffective: the agencies did not have a clear understanding of who bore which responsibility (MOWR, 1998b). The Water Resources Management Proclamation 25/2000 now assigns all the responsibilities related to water resources management, including the enforcement of compliance of water quality standards and related federal regulations, to a supervisory board under the MOWR at the federal level (MOWR, 2001b: 73). The MOWR is thus mandated to control and enforce standards, build infrastructure, develop and standardize water technologies appropriate for the country's conditions, and help train human power for local and regional water agencies.

*Inadequate collection of water bills* The recovery of water bills is very inefficient in all urban areas. Most household water meters no longer work. In Addis Ababa, for example, at least 20 percent of water meters do not function. No systematic monitoring and replacement of faulty meters exists. People deliberately tamper with water meters. Customers with broken meters are charged a flat fee for monthly water use, based on previous level of water use, regardless of the volume of water consumed (World bank, 1995: 33-36). Many water meters are not always accurate. Many customers complain about exaggerated billing due to faulty meters. Of course, under billing due to the same problem exists but is not reported (AAWSA/UNCHS, 2000: 51). In some towns, customers are charged only for the monthly water meter service if and when meters malfunction.

*Lack of community involvement* Water supply development programs rarely incorporate community participation in their plan, design, operation, and maintenance. Transparent management and accountability to the users is totally absent in the water sector of the country. The long term efficiency and maintenance of water supply facilities depends on community participation and concern. The local community must be actively involved in design, planning, construction, operation, and maintenance. Centralized or outsider efforts have failed in the past because they have not involved the community of consumers.

Community participation must involve direct decision-making and hands-on management on the part of all members of the community during all stages of development. The community must not merely receive information or ratify decisions about development: it must participate in the development process itself. Decisions should be made by those most affected by changes, that is, women in the case of the water supply and sanitation sector. In most cases women make responsible and valuable caretakers of water supply facilities because they are the primary users of water supplies. This is especially true in rural and small urban areas. Often women provide the main labor for the construction of new schemes. However, their opinion and potential role as decision makers is frequently overlooked (Water and Sanitation for Health Project (USAID), 1983: 63-65).[4]

Community participation has many advantages: it encourages a sense of responsibility, allows the provision of less costly service, ensures things are done properly and effectively, ensures the sustainability of schemes, and encourages the creative application and practice of local knowledge. As Kauzeni (1983)[5] cautions: 'Failure to inculcate the sense of ownership of the scheme in [beneficiaries'] minds through their involvement in the planning and construction phases of the water scheme discourages them from participating in the subsequent phases of operation and maintenance of the schemes.'

*Lack of skilled human power*  There is an obvious lack of skilled human power in the water sector. This has resulted in an over-dependence on foreign technicians and innumerable system failures. There are a few institutions and vocational schools that train personnel to work in the sector. Many professionals have traveled abroad for graduate and postgraduate studies. The sector has also been involved in training projects and workshops. Despite this there has not been a sufficient supply of skilled personnel to meet the demands of the sector. The job market has offered uninviting work conditions, low incentives and salaries, and has contributed to the depletion of skilled personnel working in the field (MOWR, 1998b: 45-50). Low salaries and the absence of career structure drive many technically able staff members to work in the private sector. In 1997, a 25-town water supply development and rehabilitation project survey found that water sector personnel were unskilled for their jobs, underpaid, and, thus, tended to be unmotivated about their jobs (MOWR/Ernst & Young, 1997a: 5.3).

*Over-dependence on foreign donors*  Past and present governments in Ethiopia rarely made significant public investment in the water supply sector except paying salaries. Bilateral and multilateral organizations, non-governmental organizations, and international funding agencies finance the water sector.[6] The problems in operation, maintenance, and monitoring of urban and rural water supplies are in part due to over-dependence on unsustainable expatriate funding and technical expertise. In most cases, external donor agencies rarely fund operation and maintenance or programs that are intended to improve the sustainability of services. Most technologies used in the water sector are imported from abroad and the assistance of external donors is partly responsible for this. Imported technologies are often inappropriate, especially at times of repair when acquiring spare parts can be expensive (MOWR, 1997a: 2). By and large, most water projects do not promote self-reliance; they encourage dependency instead.

## External Assistance in Water Supply and Sanitation Development

In the 1990s, various international donors and non-governmental organizations (NGOs) provided, and continue to provide, assistance to the country's water supply and sanitation sector (World Bank, 1995: 5). Urban water supply has been funded and managed with assistance from the African Development Bank, European Union, KfW (Germany), and United Nations Development Program (UNDP). Rural water supply schemes have been established with the help of Sweden, Canada, GTZ (Germany), and UNICEF. Finland, Japan, Italy, Belgium, and the Netherlands have also provided assistance. NGOs have helped provide safe water and improved sanitation in many rural areas of the country. The 1973-1975 and 1983-1985 famines in Ethiopia attracted many foreign NGOs into the country. Most NGOs assisted with food distribution and the provision of basic health, water, and sanitation services to famine-affected people. Many of them stayed in the country after the famines to help with rehabilitation and development projects. Most NGOs focused on rural areas and assisted with water supply, sanitation, and small-scale irrigation projects.

In the early 1990s, there were 70 NGOs in the country of which 40 were involved in water and sanitation activities, especially in rural areas (World Bank, 1995: 13). The NGOs focus on low-cost technologies such as spring development, borehole drilling, hand-dug well construction, and alternative technologies such as solar power pumps, windmills, and rainwater harvesting systems. Most NGOs also help with latrine and ventilated pit latrine construction. NGOs tend to localize their activities (for example, the Norwegian Church Aid in Bale and Oxfam in Afar). Large NGOs do respond to national emergencies and operate outside their original area bases. In general, NGOs have proved invaluable to the development of safe water supply and sanitation in the country, especially in rural areas. However, their efforts are usually small-scale and not enough in the face of the country's growing demand on the water sector (MOWR, 1997b: 4). Many of the water systems they help built also break down after a few years of service largely due to lack of funding for maintenance (external agencies that help build water systems will not finance maintenance and operation activities). Additionally, NGOs do not work in close collaboration. They do not communicate with each other about policies, strategies, and overall objectives; thus, redundancies, inefficiencies, and inequitable assistance between regions occur. For the most part, NGO activities are ad hoc and short-lived.

**Water Pollution**

It is difficult to assess the situation of water quality in Ethiopian water systems because of the lack of sufficient data. Existing data indicates that there has been a decline in water quality in some parts of the country (UNEP, 1992). Waterways that run through or adjacent to urban areas are heavily polluted. Cities and towns are characterized by open waterways that have become open sewers. Septic tank overflows, liquid wastes from domestic and institution sources, solid waste from markets, food stalls, and individual homes into open waterways have created a serious environmental and health problem in urban areas. The Awash basin shows an increase in salinity levels as one goes downstream. This increase is a result of irrigation run-off. The upper reaches of the river flow through Addis Ababa where industrial, domestic, and municipal pollution has greatly reduced water quality. Pollution is visible in the growth of sewage fungi and solid waste deposits of fecal and other toxic materials. Despite the gross pollution, this water is used for drinking, other domestic purposes, and livestock watering by more than 10,000 people living in the region between Lake Aba Samuel and Addis Ababa.

The Abijata-Shala National Park also shows marked decreases in water quality. Diversion of in-feeding rivers for irrigation and the extraction of soda ash have resulted in declining lake-levels and increasing salinity and alkalinity. These changes have drastically affected the ecology of the area, especially its bird life.

Lake Awasa, the shores of which form the city of Awasa (Pop. 100,000), one of the fastest growing urban areas in the country, receives heavy dosages of a variety of wastes generated by domestic, commercial, and industrial establishments. The textile industry is a major culprit. It dumps untreated toxic liquid wastes into the lake, threatening the aquatic life. Disease carrying mosquitoes breed on the shores of the lake that adjoins the city. Malaria has become more rampant than ever before and many visitors to this fast growing resort city contract the disease upon return to their domiciles in higher elevations.

The Lagadadi reservoir, which is primarily used as the principal source of water for the city of Addis Ababa, has suffered numerous problems in the last decade or so. There has been an increase in the turbidity and color of its water over the years, especially during the wet season. The reservoir is prone to siltation and the risk of eutrophication.[7] Increased iron and magnesium contents have also been recorded during the dry seasons. Algae blooms are also believed to be a perennial problem. High dissolved nutrient concentrations have been recorded. High nitrate and silicate contents are common in the months June to October because of increased run-off and tend to drop off in the months November to February when run-off decreases. The reservoir receives fertilizer by-products and herbicide from the surrounding farming communities. Bacteriological concentrations in the catchment area and reservoir are high during the rainy season (AAWSA, 1994: 1-2). Water is critical to the transmission of many diseases. In the developing world, unsafe drinking water contributes to 80 percent

**Table 3.3  Common Diseases Transmitted to Humans Through Contaminated Water**

| Type of Organism | Disease | Effects |
|---|---|---|
| Bacteria | Typhoid | Diarrhea, severe vomiting, enlarged spleen, inflamed intestine; often fatal if untreated |
| | Cholera | Diarrhea, severe vomiting, dehydration; often fatal if untreated |
| | Bacterial dysentery | Diarrhea; rarely fatal except in infants without proper treatment |
| | Enteritis | Severe stomach pain, nausea, vomiting; rarely fatal |
| Viruses | Infectious hepatitis | Fever, severe headache, loss of appetite, abdominal pain, jaundice, enlarged liver; rarely fatal but may cause permanent liver damage |
| Parasitic protozoa | Amoebic dysentery | Severe diarrhea, headache, abdominal pain, chills, fever; if not treated can cause liver abscess, bowel perforation, and death |
| | Giardiasis | Diarrhea, abdominal cramps, flatulence, belching fatigue |
| Parasitic worms | Schistosomiasis | Abdominal pain, skin rash, anemia, chronic fatigue, and chronic general ill health |

Source: G. Tyler Miller (1998), *Living in the Environment*, Belmont, CA.: Wadsworth Publishing, p. 516.

of all diseases. There are various water-related pathways by which a disease can be transmitted from one person to another (Cairncross and Feachem, 1993: 4-5; Sanders and Warford, 1976: 32) (Table 3.3). The transmission of *water-borne diseases* occurs when a person ingests the pathogen in water. Water-born diseases include cholera, typhoid, leptspirosis, giardiasis, amoebiasis, infectious hepatitis, diarrhea, and dysentery. *Water-washed diseases* occur from lack of adequate water for personal hygiene and washing. Diarrhea, bacillary dysentery, ascariasis, trichuriasis, and other water-borne diseases are among the diseases that may result. As these diseases are all fecal-oral in their transmission pathway, they are either water-borne or water-washed. *Water-based diseases* are transmitted through direct contact with organisms in water. Schistosomiasis and dracunculosis (Guinea worm) are examples of water-based diseases. Insects that breed in water spread *water-insect related diseases*. The mosquito, for example, transmits malaria, yellow fever, encephalitis, and onchocerciasis (river blindness).

**Conclusion**

Most Ethiopians do not have access to safe drinking water, despite the availability of substantial water resources. This includes about 86 percent of the rural population and nearly 20 percent of the urban population. There are shortages in both supply and quality of water. The per capita consumption is less than 10 liters per day. This amount is insufficient for health and hygiene.

Many problems plague the water sector. Investments channeled to the sector are well below what is required. Water tariffs are quite low to cover the costs of maintenance and operation. Metering, billing, and collection services are inefficient. Unaccounted-for-water losses are massive. Most water schemes are in need of rehabilitation. Water quality is often poor: many water supplies have high bacterial content and high fluorine and other minerals. In addition, the sector lacks adequate skilled personnel, incentives, accountability, and community participation in decision-making, management, and operating systems.

The federal government hopes to address the problem of water supply by working on rehabilitation projects and constructing new schemes. In 2002, it launched a 15-year water sector development plan to meet the increasing water needs of the country. The plan promises to carry out water supply study and design for 391 towns, to implement construction work for 402 towns, and to perform rehabilitation work for 112 towns. Urban water supply coverage is expected to increase from 74 percent in 2002 to 98 percent by 2016. For rural communities, the plan proposes to initiate the construction of 4,255 deep wells, 9,329 shallow wells, 27,338 hand-dug wells, 18,908 springs, and 222 sub-surface dams. In addition, the plan provides for over 2,850 water rehabilitation schemes. As a consequence, rural water supply coverage is expected to increase form 23 percent in 2002 to 71 percent in 2016 (MOWR, 2002b: 2).

**Notes**

[1] For more discussion on river basins, see Ethiopian Valleys Development Studies Authority (1992), *A Review of Water Resources of Ethiopia*, Addis Ababa: EVDSA.

[2] For a detailed early history of water supply development in Addis Ababa, see Addis Ababa Water and Sewerage Authority (1972), *Water Supply Services for the City of Addis*

*Ababa*, Addis Ababa: Addis Ababa Water and Sewerage Authority; Addis Ababa Water and Sewerage Authority (1957), *Water Supply and Sewage Services in Addis Ababa*, Addis Ababa: Addis Ababa Water and Sewerage Authority.

[3] In Ethiopia, wells up to 60 meters deep are considered shallow while those 60 meters or more are considered deep (MOWR (2001a), *Environmental Project, Component 3, National Water Supply and Sanitation Master Plan: Status Report, Volume II, Physical Resource Base*, Addis Ababa: MOWR, Annex II-1.4, p. 4).

[4] Women's role as the primary user of water and their traditional role in water supply have been taken for granted for too long. Their activities are not assigned an economic value and thus have not been given due attention. Current water supply and sanitation projects exclude the say and participation of women, despite the fact that it is women, the main users of these facilities, who determine the success or failure of the projects. Many new schemes introduce technologies that do not account for cultural and social context or know-how. Many ignore the needs of women completely (for example, the handle pumps are too heavy or are located at the wrong height for women to use effectively).

[5] Cited in Atnafe Beyene (1997), "The Role of Community Participation in the Sustainable Development of Water Resources Management," *Water*, 1, 1 (June), p.36.

[6] For the years 2000-2003, for instance, the government planned to allocate a total of 360 million Birr for the water sector of which 87 percent was expected to come from foreign sources (MOWR (2001c), *Environmental Support Project Component 3: National Water Supply and Sanitation Master Plan, Status Report, Volume III, Financial and Economic Base*, Addis Ababa, MOWR, p. 92).

[7] Eutrophication: 'Physical, chemical, and biological changes that take place after a lake, estuary, or slow-flowing stream receives inputs of plant nutrients—mostly nitrates and phosphates—from natural erosion and run-off from the surrounding basin' (Miller, 2002: G5).

Chapter 4

# Urban Sanitation and Waste Management

The United Nations Water Supply and Sanitation Decade was developed to accelerate growth in both areas. Whereas some advances have been made in the area of water supply, emphasis on sanitation was less pronounced. Ethiopia's sanitation sector leaves much to be desired. Accurate and reliable information on the status of sanitation coverage in the country is hard to come by, but a 1999 survey by the CSA indicates that only 16 percent of the population of the country had access to adequate means of sanitation. An even lower percentage of the rural population has access to sanitation services; only less than 8 percent had appropriate means of sanitation (CSA, 1999: 164).[1] By contrast, 30 percent of Kenyans, 60 percent of Ugandans, and 77 percent of Tanzanians have access to sanitation facilities (MOWR, 1998a: 28). Development of rural sanitation in Ethiopia has never been an important public policy concern.

The World Health Organization of the United Nations defines sanitation as the control of materials in the environment that threaten the health of humans.[2] Sanitation thus means more than access to pit latrines. It includes a whole process of enhancing the conditions of the living environment (both inside the home and outside), personal hygiene, as well as improving the physical infrastructure of toilet facilities, safe and adequate water supply, and the safe disposal of domestic solid waste and liquid waste.

In the last two decades, progress has been made in providing drinking water in urban areas in Ethiopia. However, few improvements have been made in the sanitation sector. Government authorities give the least attention to sanitation. The benefits from water supply are tangible whereas benefits for sanitation are not. That is why past efforts heavily emphasized improving water supply, often to the complete exclusion of sanitation. In nearly all large urban places, the number of households with access to safe water surpasses the number of those with sanitation facilities.

Nearly two-thirds of the urban population in Ethiopia use pit latrines for sanitation while close to a third defecate in open fields (Table 4.1). Less than 5 percent of the population use flush toilets. Only in large cities and towns do more than five percent of the households use flush toilets (for example, Addis Ababa, 11 percent; Mekele, 9.4 percent; and Harari, 6.8 percent).

**Table 4.1  Distribution of Households by Toilet Facility in Selected Urban Centers, 1998**

| Selected Urban Centers | Toilet Facility | | | | |
|---|---|---|---|---|---|
| | Flush Toilet (%) | Pit Latrine (%) | Container/ hh items (%) | Field/ Forest (%) | (%) |
| Mekele | 9.8 | 67.7 | 0.5 | 22 | - |
| Assayita | 0.4 | 47.3 | 1.6 | 50.5 | 0.3 |
| Gondar | 4.2 | 52.9 | 0.5 | 42 | 0.4 |
| Bahir Dar | 5.2 | 68 | 0.8 | 26 | - |
| Dessie | 4.6 | 69.3 | 0.2 | 25.9 | - |
| Jimma | 3.9 | 79.1 | 0.9 | 16 | - |
| Adama | 3.6 | 86.6 | 0.2 | 9.6 | - |
| Bishoftu | 5.1 | 80.1 | 1.4 | 13.4 | - |
| Jijiga | 1.6 | 72.7 | - | 25.6 | 0.1 |
| Asosa | 3.5 | 91.4 | - | 5.1 | - |
| Awasa | 5.4 | 88.5 | - | 5.1 | 1 |
| Gambela | 5.8 | 30.6 | 0.9 | 59.2 | 3.6 |
| Harari | 6.8 | 77.2 | 0.4 | 15.2 | 0.4 |
| Addis Ababa | 11 | 73.3 | 3.1 | 10.5 | 2.2 |
| Dire Dawa | 3.9 | 76.1 | - | 19.8 | 0.2 |
| Amhara other Urban | 0.8 | 45.5 | 0.1 | 53.4 | 0.2 |
| Oromiya other Urban | 1.3 | 64.5 | 0.2 | 33.9 | 0.1 |
| SNNPR other Urban | 1.1 | 76.4 | 0.1 | 20.9 | 1.5 |
| Other Urban | 4.8 | 37.1 | 0.6 | 57.4 | 0.2 |

Source: Central Statistical Authority (1999), *Reported on the 1998 Welfare Monitoring Survey*, Statistical Bulletin No. 224, Addis Ababa: CSA, p. 181.

**Sanitation in Addis Ababa**

Residents of Addis Ababa face the most insidious environmental problems, largely due to poorly developed environmental infrastructure. In 1998, 11 percent of the households in Addis Ababa had private flush toilets, 73 percent had private or shared pit latrines, and 16 percent of the households had no toilet facilities of any kind (CSA, 1999: 181). In the peripheral areas of the city, 40 percent or more households had no sanitation facilities. City residents who lack latrines must defecate in nearby open grounds hidden from public view or walk longer distances to ravines, ditches, or wooded areas.

Addis Ababa has inadequate, hygienically deplorable sanitation facilities. Most shared latrines are unventilated, overused, unlined, collapsed, and overflowing. The unlit pit superstructures are old and dangerous: wooden planks covering pits are unstable and fouled. On the average, 34 people share a pit latrine. Over 70 percent of the latrines are shared with 20 or more users. The densely populated central areas have the highest percentage of households (50-60 percent) using shared pits. Many of the *Kebeles* in this area have 70-75 percent of their population use these latrines. One of the most poorly served, densely populated areas in the city is the Merkato. Over 60 percent of the residents of the Merkato area use shared latrines and 25 percent have no toilet facilities. The shared latrines, crowded in between houses in collapsing superstructures, are overused and overflowing. Cleaning tankers cannot access them. Because of the lack of toilet facilities, roads often overflow with excreta and refuse.

In densely populated districts, the lack of space often forces large numbers of households to share the same latrine. Not surprisingly, *kebele* housings have the worst sanitation service. Tenants' lack of ownership to such dwelling provides little or no incentive to keep them clean, make improvements in existing facilities, or build new ones. This also leads to misuse and unhygienic and unsanitary conditions.

Private pit latrines are better constructed and maintained than shared ones. Most are unventilated (though some newer low-income housing now have ventilated pit latrines) and are made of zinc sheets, *chika*, or other scrap materials. Most private latrines have fewer than 10 users. The proportion of privately owned houses is greater in the outskirts of the city and, thus, private latrines constitute 20-25 percent of the sanitation used. By contrast, less than 10 percent of the sanitation consists of private pit latrines. Income levels and density influence the use of private latrines. The central areas are congested and house ownership issues are more likely to restrict construction of private latrines. The use of this form of sanitation is about 15 percent in the areas between the periphery and the center of the city.

About 11 percent of the households of greater financial means in the city have private or shared flush toilets. Most of these toilets are not hooked to the main sewer network. Septic tanks, cesspools and open waterways are used for discharging sludge.

There are 72 public toilets scattered in the city. Most are located in central commercial areas and a few are located in the peripheral commercial areas. They are extensively used and only men have access to them as they have no separate sections for women. All are maintained by the Municipality and are designed as squatting plates with flush systems. Most of them are dreadfully grubby. Only one of these toilets is connected to the sewer systems. Thirty-eight of them have septic tanks that are connected to storm water drains and streams; the rest have no septic outlet and are emptied about 2 times a week by tankers. These toilets are estimated to have 1,800 users/toilet/day. Many of the cisterns are broken and flush continuously. Because the Municipality receives water at a reduced rate of 10 cents per cubic meter, there is no real incentive to attend to repairs. Many are not easy and safe to use, especially for children, the elderly, and disabled people as they are too poorly lit. They have no provision for anal cleansing and hand washing. Twenty-five percent of the toilets are estimated by the Municipality to be non-functional at any time due to problems with maintenance. At any rate, the maintenance and cleaning of public toilets are so poor that many people avoid using them.

### Liquid Waste Management in Urban Areas

Liquid waste or wastewater is water borne. Individual households, shops, and institutions like hospital, industries, and other establishments generate it. Human excreta (sewage) and industrial waste form the principal components of liquid waste. AAWSA manages Addis Ababa's residential and institutional liquid waste. The Environmental Health Department (under the regional administration) handles the collection and disposal of liquid waste from public toilets. Municipal governments and local administrations manage the liquid waste of all other urban areas.

Liquid waste disposal management in the country's urban areas is piteously poor. With less than one-half of the urban population served with sanitation facilities, defecation in open fields is the normal practice. Liquid waste overflows from pit latrines and septic tanks and is often disposed without treatment into open streams and waterways. Industries also discharge liquid waste directly into streams. They are able to do so because there are no stringent laws and regulations to prevent them from doing so. Methods of liquid waste disposal are generally found to be inappropriate and harmful to the living and working environment of urban areas.

Urban area municipalities are the sole government caretakers of liquid waste management. They plan and administer social, environmental, health, education and other infrastructural services and are thus over burdened with work. Consequently, the liquid waste management program is given a low priority. Moreover, municipal departments in charge of liquid waste lack effective planning and management systems, adequate finance, equipment and skilled human power, and enforcement, monitoring and evaluation systems.

## Liquid Waste Management in Addis Ababa

About 72 percent of the population of Addis Ababa use pit latrines and septic tanks for their household liquid waste disposal. Of this, 12.5 percent used septic tanks, 39 percent use shared pit latrines, and 21 percent use private pit latrines. Only 1.2 percent of the population is connected to sewer lines. About 27 percent of the population dispose of their liquid waste in ditches adjacent to the house, but the practice obviously damages health and the environment.

The per capita per year sludge production in the city is 150 liters for shared pit latrines, 100 liters for septic tanks, and 100 liters for private pit latrines. The per capita per day liquid waste for sewer lines was estimated to be 150 liters. Thus, the total liquid waste produced in Addis Ababa in 1996 was estimated at 6,700 cubic meters/day. Approximately 800 cubic meters/day is sludge from pit latrines and septic tanks and the remaining 5,900 cubic meters/day was liquid waste carried along sewer lines.

AAWSA manages the city's liquid waste disposal, except for the public toilets that are in the charge of the Region 14 Health Bureau. AAWSA has 25 vacuum tanker trucks used for sludge collection and Region 14 has 15 vacuum tanks. More than half of the vacuum tanks are old and many of them are usually in the repair shop at any given time. Only 60 percent of the total sludge produced in 1996 was collected and disposed off. Only 45 percent of the city's population was provided with liquid waste collection services.

Households have to request sludge tankers from AAWSA well ahead of time, as a waiting period of 4 weeks (in some cases up to 12 weeks) is common. Though some customers (such as institutions, hospitals, associations, and those with over-flowing pits) are supposed to receive priority service, many organizations and individuals are made to wait for long periods of time. Preferential treatment to customers who can pay 'extra incentive money' to sanitation workers is a common practice. Many customers have to wait 3 to 6 months and most of the latrines are over-flown when finally serviced. Emptying pit latrines and septic tanks used to cost between 25 and 30 Birr, but these fees were increased to 49 Birr in 1991, first, so that the authority could cope with its shortage of vacuum tankers and second, to cover the costs of additional weekend and evening services and worker over-time. Though the fees have increased, evening services are rare. Customers located in peripheral areas of the city (districts like Kotebe in the east and Gulale in the west, for example) are charged 75 Birr.

The private sector has increasingly involved itself in the sanitation sector. Three private enterprises clean and empty pit latrines and septic tanks in the city. Gebre Egziabher and Efrem Bekele have one truck each with a capacity of 8 cubic meters. They each charge 130 Birr per trip or 16.25 Birr per cubic meter for their service. The Universal Trading and Commission Agency has five trucks with a capacity of 8 cubic meters. It charges 150 Birr per trip or 18.25 per cubic meter. Some NGOs also have vacuum tankers that they rent out to users within their project communities (UNDP/World Bank, 1997: 56-57).

Many problems plague the sludge collection system. The current vacuum tanker fleet cannot meet the demands of the city. A perpetual shortage of spare parts keeps many vacuum tankers out service. The waiting period between servicing is quite long and the repair system is prone to corruption. The existing vacuum tankers cannot access an estimated one-half of the latrines in the city. Tankers are 2.5 meters wide, while paths to latrines are often less than a meter wide. Many latrines are more than 35 meters away from the nearest vehicular road; thus, the 40 meters hoses of the tankers cannot reach them. The length of the tankers (7 meters) makes them less maneuverable. A build-up of solids and rubbish in pits can hinder latrine-emptying operations by causing blockages in the tanker hoses. Because the hoses do not have grits at their mouths, solid materials (for example, rags, glass, cans, bones, plastic, sheets, shoes and paper) can enter the hoses and clog them. Clogged hoses slow the process of latrine emptying: oftentimes clogged hoses cannot complete the job. Unemptied latrines are either abandoned or the sludge is diverged to the nearest storm drain. When latrines are full or cannot be emptied, people either continue to use the full latrine and the surrounding area, use a bucket and empty it in nearby ditches, use open grounds, or use a public latrine.

High-density areas are by far the most poorly served in terms of sanitation facilities. In these areas there are not enough latrines for the general populace. This results in over-use and existing conditions are worsened by the poor construction, design, and maintenance of these latrines. Maintenance requires access to latrines by tankers, tanker availability and frequency, facilities for sludge, and storm water disposal—all of which are inadequate or lacking. Poor maintenance results in overflowing latrines and flooding, especially during the rainy season. New latrines are not constructed for several reasons, including the lack of space and the cost. Moreover, many people do not own their housing and they have little motivation to improve it.

People simply ignore sludge emptied in ditches and storm drains. Broken sewage lines are not repaired on time even under circumstances that threaten public health. In March 1998, for instance, a sewage line broke near the Ginfile Bridge and released on the street tens of cubic meters of human waste day in and day out for several weeks, attracting swarms of disease-bearing insects. Tens of thousands of people cross this bridge every day. The stench was so nauseating that people took the long detour to get to the other side of the city. Others who passed through this bridge had to hold their breath and walk fast or run to avoid the loathsome smell and the risk of being splashed with filth by fast driven cars. Only after several weeks of obnoxious environmental pollution and public outcry did the water and sanitation agency fix the broken sewage line.

Sewer lines are almost non-existent in urban areas. Addis Ababa has a 110-kilometer long sewer line that serves only 7 percent of the city's population (Malifu, 2001: 142). Three percent of the populations of the other major towns use septic tanks and about 47 percent use pit latrines. Fifty percent of the populations have no sanitation facilities at all and, as a result, sludge collected is simply dumped into nearby river ways. The banks of all the rivers that criss-cross many urban areas, including those in the city of Addis Ababa, are covered with filth of the worst kind.

Addis Ababa's sewerage network was constructed between 1976 and 1982 and was expanded between 1984 and 1990. The sewer system was laid down in areas already connected to a water supply network. These areas included residential neighborhoods of Lideta (in the west), Bole (in the east), and in the commercial and business districts in the city center. The design incorporates three main interceptors in the city's south. The Lideta and Western interceptors (for the Gofa Safar, Mekanisa and Lideta areas) meet at the junction of the branches of the Akaki River in Mekanisa. The west bank of the Qabbana River and Bole and Urael areas are served by the eastern interceptor, which would also have an extension built out to Abiyot square and the University. The interceptors from the west and east converge as a single pipe on the left bank of the Akaki River and are treated at a plant further downstream.

The system has been designed to accommodate liquid waste generated at 150 liters per capita per day for a population of 200,000 people. It currently has a lower flow because only 1,500 households are connected to the system. Polluted water from the little Akaki River is diverted into the Western collector to ensure the minimum flow required for the operation of the treatment plant. The water flow is primarily gravity induced, except in the Lideta Area where there is a pumping station. Household connection to the sewer system is charged for.

Liquid waste from the sewer system and sludge from pit latrines and septic tanks are treated at the Kaliti treatment plant, located south of the city on the left bank of the Little Akaki River. The design of the treatment plant incorporates earth ponds where liquid waste can be oxidized and stabilized for relatively long retention periods. Increase in quantities of liquid waste carried through the system can be countered by installation of mechanical aeration and settling tanks upstream of the stabilization ponds. The system is built to handle 7,600 cubic meters sewage/day (biochemical oxygen demand load of 3,500 kilogram/day). An independent stabilization plant constructed to process septic tank and pit latrine sludge can handle 250 cubic meters of vacuum tanker material/day (biochemical oxygen demand load of 2,500 kilograms/day).

The existing treatment plant suffers from many deficiencies. The liquid waste treatment plant works at less than 20 percent of its design capacity. The total daily inflow rarely exceeds 1,500 cubic meters as few households are connected to the sewage system. The septic tank and pit latrine sludge treatment plant were found to be unsatisfactory in their drying and settlement functions and thus have been completely abandoned. The sludge that is brought to the plant is being dumped in the discharge channel connected to the Little Akaki River. The river is already highly polluted by the time it flows by Kaliti and the added liquid waste degrades the environment further. Lake Aba Samuel downstream now suffers from nitrification and the growth of water hyacinth. The Awash River, one of the country's main rivers, joins up with the lake and is threatened by pollution as well. The pollution of these waterways causes a serious health risk to all users.

Manufacturing industries, hotels, and hospitals are major pollutants in the city. There are almost no restrictions imposed on their disposal of liquid waste. All discharge waste into the rivers in the city. Some organizations have treatment plants, but these are under-sized and many are non-operational. The main industrial polluters are tanneries, breweries, chemical and food processing plants, and textile businesses (more on these in Chapter 5).

Addis Ababa has a storm water drainage system that extends under most of the city's paved roads. The system consists of underground pipelines and open earth ditches that empty into the south flowing rivers and waterways of the city. Most of the network is focused at the city center along the main roads. However, the storm sewers do not drain nearly two-thirds of the roads. Most of the storm sewer systems that are supposed to be functional are in a state of disrepair. At least two-thirds of the network requires rehabilitation. Many are silted up and, as a consequence, their capacity to cope with rain run-off is greatly reduced. Many street inlets are non-functional, clogged with solid wastes, or destroyed by traffic.

The geography of the city would have helped reduce flooding had the existing system of run-off discharge properly functioned and had people not settled in flood prone areas. Addis Ababa is situated downstream of mountainous terrain with steep gradients, many rivers, and a large number of natural outlets. Water stagnation should not have posed a problem for much of the city because surface run-offs after the rains are usually of short duration and high velocity. However, the paucity of vegetation in the catchment areas, the inadequacy and/or the collapse of the storm drainage network, increased settlements near riverbanks and in steep areas, and expanded road networks have all increased the potential damage caused by flooding.

The effects of large volumes of run-off are seen in damages done to roads, road shoulders, embankments, and the foundations of many buildings. Streets inundated by floodwater make movements in the city torturous and hazardous. An editorial in *The Addis Tribune* (2000) characterized the situation: 'It is better that drivers and pedestrians alike use canoes to go about town.... After a torrential rain, almost all the roads in the city seem to be meandering rivers in which children splash gleefully, daring drivers dash about as if they own motor-boats. Poor pedestrians jump and run or coil at a corner at a sight of one of those cars. It is a common scene to see some unwary passerby taking a shower out in the open but in full regalia'.[3]

The storm water drainage system disposes of domestic, industrial, commercial, private, and public liquid waste in the city, either directly or from septic tanks or cesspools. The system creates major environmental hazard.

Sullage water flows along open ditches beside roads and often overflows. Waterways and rivers are prone to sewer pollution, especially in areas that are heavily settled, and during the dry season when the flow is primarily liquid waste. Mosquitoes and flies breed in these areas, increasing health hazards. Pipelines are often clogged. Gases released from decomposing sewers corrode the pipelines and the strong stench of rotting sewer is also a nuisance. There is no comprehensive legislation for liquid waste disposal and effluent discharge into ditches and natural waterways. Though recent laws call for attempts to curb pollution, as yet, no regulations enforce these laws.

Gondar is the only other major city with a piped sewerage system, built by the Italians in 1940. The city has two separate lines that serve the center of town. They empty into the Bisnit and Kaha river valleys. The pipes are small in diameter and clog often (MOWR, 1997b: 4.43). Municipal governments in other major cities and towns use vacuum tankers to desludge septic tanks, pit latrines, and public toilets. However, shortages of vacuum tankers and poor access to many high-density neighborhood facilities limit their ability to service only a few urban households. None of the major urban areas has a liquid waste treatment facility; hence, vacuum tankers simply dispose of the sludge at the outskirts of town roadsides or river channels.

A 1992/1993 Master Plan for the development of liquid waste facilities in Addis Ababa outlines strategies to ensure improved sanitation by 2015 (AAWSA, 1996). The proposal includes the expansion of the existing sewerage system, improvements in the collection, treatment and disposal of fecal sludge, the development of a public health and hygiene education campaign, and strategies to control industrial and institutional liquid waste. The African Development Bank financed the plan.

The Master Plan envisions the development of new systems for the safe treatment and disposal of fecal sludge, efficiency improvement in fecal collection and disposal, and a reduction in health risks generated from pollution by the Kaliti treatment plant discharging into the Little Akaki River and in Lake Aba Samuel. Efficiency improvements in sludge collection services will include: 1) the introduction of two new types of smaller collection units (2 cubic meters) to meet demands of pit latrines that are inaccessible to the larger (8 cubic meters) vacuum tankers; 2) the construction of five sludge transfer stations in the city where the small trucks can discharge their sludge into larger (6 cubic meters) tanker trailers that could be towed to the treatment plant by 8 cubic meters trucks; 3) the addition of vacuum trucks with self-contained water tanks and jet facilities that can help remove more dense sludge from latrines; 4) the addition of large (14 cubic meters) trucks for use in areas with good access to help increase cost efficiency; and 5) the provision of repair facilities—equipment and training—for the proper maintenance of the fleet.

The Master Plan also recommends land application (deep ploughing and forest development) to treat and dispose of sludge until 2009 and sludge pasteurization thereafter. The Plan suggests the construction of drying sludge bed and sludge burying until land application is fully developed. The construction of temporary lagoons is recommended to help contain sludge during the wet months of June to September when vehicle accessibility and effective sludge drying are hindered by the rainy season. The Plan identifies the Kotebe area as the location for the development of forests and the construction of sludge drying beds and lagoons. The pasteurization plant is planned as an addition to the existing facilities at Kaliti. The plan also envisions the modification of existing but abandoned units at the current treatment plant. These improvements will treat an estimated 297,840 cubic meters of sludge in the year 2005 and 536,112 cubic meters in the year 2015. The implementation of the Master Plan has been slow, however. By the end of the 1990s, tangible sanitation improvements were yet to be evident in the city. Until the capital establishes an effective sanitary system sewage will continue to present a major health threat and an aesthetic nightmare.

## Liquid Waste, Health, and the Environment

Liquid waste degrades the environment as it washes into streams and pollutes water resources. Industries also discharge liquid waste directly into streams: no stringent laws or regulations prevent them from doing so. Methods of liquid waste disposal are generally found to be inappropriate and harmful to the living and working environment of urban and rural areas.

Liquid waste comprises a major public health hazard in urban areas, especially in high-density neighborhoods with few sanitation facilities and frequent overflowing septic or pit latrines. However, the public health risk is not geographically limited to the areas where such overflowing occurs, as the sludge can spread over a wide area and affect a much larger population. The lack of sanitation services thus adversely affects the entire community: indiscriminate disposal of waste presents a community hazard.

Human excreta, a principal component of liquid waste, is a critical path for diseases. Children form the most direct route from excrement to illness when they play with fecal-contaminated soil and other materials and then place their hands in their mouth or on their food. Less direct routes include vectors such as flies and cockroaches and polluted water sources. Flies breed in vast number in human and animal excreta, sewage sludge, and waste dumps. Because they live in close proximity to humans, flies probably spread the greatest number of diseases to people. Highly mobile, they move within and between houses and infect exposed food, liquids, and utensils. Flies are frequent transmitters of such diseases as cholera, various types of diarrhea, dysentery and many food-born diseases. Children suffer the greatest number of deaths due to these diseases.

Most people in the country depend primarily on traditional water sources (lakes, rivers, streams, and groundwater) that are often polluted by untreated liquid wastes. The prevalence of intestinal parasites and diseases (diarrhea, cholera, typhoid, hepatitis) among the population is a direct result of this.

Childhood diarrhea is a leading cause of morbidity and mortality, especially among children younger than five years old. Women, as water carriers and household caretakers, are the most vulnerable to water related diseases.

Untreated sewage is detrimental to water ecosystems in and around the city. Sewage contains nitrates, phosphates, and numerous other materials essential in animal and plant metabolism. But at excessively high concentrations, these nutrients can ruin aquatic ecosystems. The most common consequence is the blooming of a huge population of algae. The overgrowth of algae throttles other aquatic life by using up much of the oxygen in the water. Algae also block out sunlight. In the end, fish and other organisms in the water rapidly disappear.

Eight in ten food and drink establishments in urban areas are considered major sources of contamination (Region 14 Administrative Health Bureau, 1997: 132). Lack of proper sanitation facilities, unsafe food handling practices, poor personal hygiene, indiscriminate street food vending practices, and the absence of standardized and sustained hygiene education all contribute to the high rate of food related diseases that affect a large proportion of the urban population. For instance, many of the tens of thousands of food and drink establishments operating in Addis Ababa are considered extremely potent sources of disease and environmental pollution. Of the 18,000 or so inspections of food and drink establishments that took place in Addis Ababa in 1995, nearly 70 percent violated one or more of the hygiene codes of the city and were deemed unfit to operate (Kumie, 1997: 116). Many of the establishments suffered from inadequate sanitation, inadequate waste collection, inadequate facilities to prepare, cook, and store, and inadequate personal hygiene (for example, inadequate hand washing after defecation) by employees handling food. Many street food vendors also sell unhygienic food and greatly exacerbate the risk of disease. The rates of illness and death from food-borne pathogens in urban areas are impossible to estimate due to the absence of surveillance and reporting systems. However, studies conducted elsewhere in the developing nations show that overcrowded conditions, inadequate water supplies and sanitations services, and inadequate facilities for processing, cooking and storing food greatly increase the risk of food contamination (McGranahan, 1991: 24).

Most food and drink establishments in the city invariably dump their wastes directly into the environment and water resources. By law, sanitary or environmental health inspectors from the city's health bureau must monitor the health conditions of all food and drink establishments, public institutions like schools and health institutes, water points, food factories, and outlets. When establishments fail to comply with the existing sanitary laws, standards, and acceptable practices, inspectors have the legal authority to take action against violators. They may issue written admonition demanding immediate corrective measures or suspend or revoke a license permit for disregarding the forewarning. But many establishments avoid punishment or buy security from prosecution by corrupting inspection authorities. Additionally, thousands of small-scale food and drink establishments operate outside the law without obtaining permits, paying taxes, or applying for the authorization required by law. Many of these establishments bribe authorities to turn a blind eye. Nearly all of these establishments lack adequate waste disposal facilities and cleanliness.

## Solid Waste Management in Urban Areas: Sources, Types, and Composition

Solid waste is non-liquid, non-gaseous disposed-of matter (Miller, 1998: 565). Most human activities create waste. Urban areas produce large volumes of solid waste. The quantity of solid waste generation generally depends on the size of population and the level of urban household incomes. The higher the income the larger the amount of solid waste produced as growing incomes facilitate higher consumption of goods and services. Addis Ababa and other urban areas generate a variety of solid wastes. Solid wastes generated from urban activities include those from households, street sweeping, commercial, industrial, and institutional categories. Typical wastes generated in each of these activities include the following:

*Residential solid wastes* Households generate about 65 percent of Addis Ababa's solid wastes (MOH, 1996: 53). Residential solid wastes contain putrescible (rapidly decomposing) animal and vegetable matters resulting from the handling, preparation, cooking, and consumption of foods, paper, cardboard, textiles, leather, wood, tin cans, yard wastes, grass, ash, and dirt. In addition, there are bulky household wastes, especially from well-to-do households, such as old furniture, appliances, and electronic gadgets. Residential wastes may include hazardous materials such as spent batteries, chlorine bleach, outdated medicines and medical bottles, detergents, paint products, and insecticides. The volume or rate of residential solid waste produced varies from day to day, month to month, season to season, and household to household. Market days yield more waste than other days of the week. Residential waste rates usually peak during holiday seasons (such as the New Year, Christian holidays like Meskel, Christmas, and Easter, and Muslim holidays like Maulid, Eid ul-Fitr/Ramadan, and Eid ul-Adaha/Arefa). People generate more waste during the dry season than during the wet season. High-income households generate twice as much waste as low-income households, because low-income households consume fewer resources and tend to recycle. The composition of household solid waste in Addis Ababa and other major cities has dramatically changed over the last ten years: there has been considerable increase in the proportion of plastics, paper, packaging materials, and alkaline batteries—a potential source of toxic metals.

*Commercial solid wastes* The sources of commercial solid wastes include such establishments as restaurants, stores, markets, office buildings, hotels, print shops, service stations, and auto repair garages. Commercial solid waste may include paper of all types, wood, plastic of all types, cardboard, food, rubber, glass, ferrous metals, textiles, worn out furniture, and used or broken office gadgets. Commercial establishments produce hazardous wastes related to the services they provide: for example, solvent from photographic and dry cleaning shops, cleaning solvents from auto repair garages, ink from printing shops, and paints and thinners from hardware shops. The volume of commercial solid wastes produced in the city is estimated at 10 percent.

*Institutional solid waste*   Institutional solid wastes are estimated at about 6 percent of all city wastes. Institutional sources of solid wastes include schools, hospitals, government offices, and prisons. In addition to the types of solid waste generated by commercial establishments, institutional wastes include the hazardous materials of healthcare establishments such as discarded syringes, needles, discarded pharmaceuticals, soiled wastes, anatomical wastes, cultures, and chemical wastes like the formaldehyde and phenols in disinfectants and the mercury in thermometers and blood pressure gauges. The waste may also include inorganic chemicals such as sulfuric, hydrochloric, nitric, and chromic acids, sodium hydroxide and ammonia solutions; oxidants such as potassium permanganate, and potassium dichromate, and reducing agents, such as sodium bisulfite and sodium sulfite (Pruss et. a., 1999: 6).

*Street solid wastes*   Street wastes—estimated at 13 percent—include street sweepings, roadside litter, haphazardly thrown residential wastes, wastes from municipal litter containers and increasing volumes of paper, leaves, animal wastes, plastics, wrappings, tins, rubber, leather, and broken glasses.

*Industrial solid waste*   Industrial solid wastes amount to about 6 percent of the city's waste. Sources and types of solid wastes generated by industries are many: metals, plastics, rubber, paper, scrap materials, cloth, glass, leather, clay, ashes, and ceramics (see also discussion under industrial pollution in Chapter 6). Many of the industrial solid wastes are considered hazardous because they may contain toxic substances. Hazardous or toxic wastes are materials that threaten living organisms. The dangers of toxic wastes include corrosivity, explosivity, flammability, ignitability, and reactivity while health-related dangers include carcinogenicity, infectivity, irritant (allergy response), mutagenicity, toxicity, and radioactivity (Tchobanoglous, et. al., 1993: 100). Leading producers of hazardous industrial wastes include chemical, leather, metal, and printing manufacturers.

*Construction solid wastes*   It is difficult to quantify wastes from the construction and repair of public and commercial buildings; however, they are substantial. The composition of such wastes may include concrete, stones, bricks, blocks, scrap wood, metals, plastics, broken glasses, plumbing and electrical parts, and dirt. Wastes from torn-down houses or buildings, crumbling streets and sidewalks, and other run-down structures also contribute to the growing volume of solid wastes in the city.

*Agricultural solid wastes* Agricultural wastes and residues resulting from various urban agricultural activities—such as the harvesting of vegetables along some of the city's rivers, the production of dairy products, and the production of small animals for slaughter—are also on the increase. These wastes are indiscriminately thrown along roadsides and waterways and in open fields. No data are available on the amounts of solid wastes generated from urban agricultural activities.

The wastes produced by residential, commercial and institutional sectors are considered non-hazardous (even though they may contain some hazardous materials) while most of those produced by industries are toxic and thus considered hazardous. The residential and commercial portion of solid waste makes up about 75 percent of the wastes generated in Addis Ababa. About three-fifths of the city's waste constitutes organic matters, mainly kitchen wastes and discarded vegetables and spoiled food wastes and grasses (AACG, 2001: 58).

It is extremely difficult to know the exact quantity of solid waste produced in the city. It may be possible to measure the amount of wastes collected by the municipality. However, the measurement would not represent the actual amount generated because not all wastes are collected. Most wastes are simply thrown in ditches, along roadside, and in waterways; or they are burned at locations where they are produced; or scavengers recycle them; or they are composted or discharged in pit latrines and municipal sewers. All these confounding factors hinder the assessment of the true rate of waste generation in the city. Most solid waste generation rates reported by municipal agencies and consultants are based on expert estimates, not the actual amount generated. On the basis of such estimates, Addis Ababa is said to produce about 0.21 kilogram of solid waste per person per day. This is a low per capita waste yield compared to other major cities in developing countries such as Jakarta, Indonesia (0.60 kilogram/person/day), Tunis, Tunisia (0.56), Calcutta, India (0.51), Cairo, Egypt (0.50), and Lagos, Nigeria (0.46). But the volume of solid wastes produced per square meter is comparable—370 kilograms for Addis Ababa, 500 for Calcutta, 250 for Lagos and Jakarta, and 175 for Tunis (MOH, 1996: 54-56).

## Solid Waste Collection

Solid waste storage and hauling containers used in homes come in different shapes and sizes and are made from a variety of materials. The most common types of containers used in urban households include plastic bags, cardboard boxes, bamboo or straw baskets, and plastic or metal containers (mostly used by affluent households).

Uncollected solid waste is one of the most visible environmental problems in urban areas. Urban municipalities are generally responsible for solid waste management, but few have adequate capacity to deal with the waste that they produce. In 1996, Addis Ababa had only 25 side load trucks and 38 lift trucks to collect and transport solid waste and 4 compactors and 4 bulldozers operating at the municipal dumpsite. Most of these vehicles are obsolete (most with 15 or more years of service) and suffer from neglect and lack of care and maintenance. The overall efficiency of these equipments in 1996 was estimated at 60 percent (MOH, 1996: 57-58). The city's vehicle maintenance shop has insufficient mechanics to do repair jobs. That is not the only problem, however. The shop is unable to perform even simple things like changing the filters or oil of the trucks. Trucks are kept off the road unnecessarily for hours at a time to cool the engine due to failure to replace plastic header tanks.

In 1997, there were 516 open-top metallic containers (each with a unit volume capacity of 8 cubic meters) distributed across the city for temporary storage of solid wastes before final pick-up by lift trucks. These are roadside containers, into which households throw their solid wastes for collection, that can then be towed away when full. Three-fourths of these containers serve residential communities without any charge while the remaining containers serve large establishments for a charge of 11 Birr per cubic meter of solid waste, a charge that covers only one-half of what it would cost the municipality per cubic meter of waste (Addis Ababa City Government, 2001: 56). The containers serve as dumping points for 85 percent of the solid wastes collected; door-to-door service handles the remainder. Each container serves an average resident population of 6,000 or 1,200 households. This ratio is woefully inadequate to temporarily store the increasing volume of waste generated in the city. According to sanitation experts, a desirable ratio is one container per 2,300 population or 460 households (Region 14 Administration Health Bureau, 1997: 15-16). Many of these containers are not well placed to offer easy access to users. In many neighborhoods, users have to walk between 0.5 and 1 kilometer to reach a container, and as a result, much waste is spilt along the way. In many locations there is only one container when, in fact, more are needed to service the huge volume of waste generated. Consequently, over-flowing containers present a common sight in the city. A one-day random survey of 89 waste storage containers in Addis Ababa in March 1998 showed that 67 (75 percent) of the containers were full or over-flowing. When this happens, the residents living nearby endure much of the environmental burden.

In 2003, the amount of solid waste generated in the city daily was estimated at 1,842 cubic meters (*Reporter*, 2003). Not all of this waste gets collected and transported to the city's dumpsite due to resource problems and inefficient infrastructure. The municipal administration says it collects about 60 percent of the waste generated in the city; the rest is dumped by the roadside, on open space, or into nearby rivers and streams, thus creating a depressing urban environment. Some households with enough space in their compound burn their wastes. Whereas this practice reduces the volume of the waste and prevents the breeding of insects and rodents, it often leads to wind-blown debris and smoke that becomes a nuisance to nearby residents and passers by.

Obviously, the quantity and type of solid wastes generated in households vary with income. High-income households generate more wastes that may include not only food byproducts but also paper, glass, metal, plastics, and worn-out and broken electronics and other old household appliances. Low-income household, on the other hand, not only produce fewer wastes, but they also recycle most reusable items. For example, used bottles, cans, plastic bags, and plastic containers are cleaned out and reused as many times as they can last and for many different purposes. The problem of collecting wastes, however, tends to be more severe in poor households and neighborhoods. In most cases, waste collection service does not extend to poor neighborhoods. If it does, collection is infrequent. In any case, most streets in poor and densely populated neighborhoods are too narrow to allow service vehicles to collect wastes.

A growing number of private businesses are involved in solid waste collection in Addis Ababa. Six private firms operate in *Wereda* 17, serving about 5,000 households. Two firms operate in *Wereda* 23, serving about 400 households. Six other private firms also operate in different *weredas* throughout the city (*Reporter*, 2003).

Other urban areas in Ethiopia also suffer from poor solid waste collection systems. Few urban areas have door-to-door solid waste collections services. Collection equipment, such as lifters, side load trucks, and hydraulic compactors, are usually not available to most urban areas. As a result, most solid wastes are burnt and/or dumped on any available empty field or in nearby ditches or rivers or simply along roads (Table 4.2). In 1998, for example, Mekele and Dire Dawa managed to collect only 47 percent of their solid wastes while Bahir Dar collected 39 percent; Jijiga, 31 percent; Gondar, 28 percent; Dessie, 25 percent; and Harari, 20 percent.[4]

Municipal public health legislation has included laws and regulations that emphasize proper waste handling, collection, storage, and disposal. In the 1950s, the Ministry of Public Health mandated that individuals, shops, and other establishments deposit solid wastes daily in the municipal garbage bins or stored in private containers until the municipal sanitation department collects it. The rules prohibited (under penalty) the indiscriminate disposal of putrid materials that breed flies and rodents. The more recent Public Health Proclamation No. 200/2000 reinforces these rules. However, all the rules in the books continue to be largely unheeded due to a general lack of enforcement, and inefficient municipal collection and disposal services exacerbate the situation.[5]

Solid waste is one of the most omnipresent environmental insults in urban areas. As serious as the public health aspects of solid wastes are, health seems not to constitute a major concern when municipal governments allocate budgets to various municipal services. The Addis Ababa municipality spends only 5 percent of its revenue on the city's solid waste management. Municipalities in many developing countries spend anywhere between 20 and 50 percent of their operating budget for these services (Bekele, 1999). Uncollected solid waste is not only unsightly but also seriously threatens health. Twenty-two human diseases stem from haphazard solid waste disposal (Vesilind, et. al., 1990: 159). A number of surveys suggest that 50-70 percent of diseases in urban Ethiopia are linked

**Table 4.2  Distribution of Households by Method of Solid Waste Disposal in Selected Urban Centers, 1998**

| Selected Urban Centers | Solid Waste Disposal Facility | | | |
|---|---|---|---|---|
| | Waste disposal vehicle/container (%) | Buried/burnt (%) | Dump in open space (%) | Others (%) |
| Mekele | 46.8 | 18.5 | 29.8 | 4.9 |
| Assayita | 4.9 | 3.2 | 63.2 | 28.7 |
| Gondar | 28.5 | 7.5 | 60.7 | 3.3 |
| Bahir Dar | 39.1 | 11.5 | 38.8 | 10.6 |
| Dessie | 25 | 11.2 | 45.7 | 18.1 |
| Jimma | 14.4 | 21.3 | 57.2 | 7 |
| Adama | 4.6 | 36 | 28.9 | 30.5 |
| Bishoftu | 5 | 34.3 | 41.6 | 19.1 |
| Jijiga | 30.9 | 8.6 | 43.1 | 17.4 |
| Asosa | 0.7 | 57.9 | 34.5 | 7 |
| Awasa | 2.2 | 55 | 27.6 | 15.1 |
| Gambella | - | 11.7 | 77.3 | 11 |
| Harari | 20.1 | 36.6 | 34.9 | 8.4 |
| Addis Ababa | 56.4 | 5.1 | 25.8 | 12.8 |
| Dire Dawa | 46.5 | 11 | 37.4 | 5.1 |
| Amhara other Urban | 1.6 | 18.4 | 72.9 | 7.1 |
| Oromiya other Urban | 2.6 | 27.3 | 57.4 | 16.2 |
| SNNPR other Urban | 3.7 | 36.2 | 50.9 | 9.3 |
| Other Urban | 10.4 | 31.8 | 55.7 | 2.1 |

Source: Central Statistical Authority (1999), *Reported on the 1998 Welfare Monitoring Survey*, Statistical Bulletin No. 224, Addis Ababa: CSA, p. 184.

directly to poor sanitation; furthermore, solid waste is a primary pathway in the transmission of diseases (MOH, 1996: 11). Biological vectors, such as rats and flies, thrive in solid wastes and have the potential to directly or indirectly transmit disease agents to humans. Vectors burgeon around or within the solid waste dumps, which are often located close to houses or on open spaces. The vectors then become sources of transmission. People in poor and densely populated neighborhoods suffer most, particularly children. Their activities in streets in close proximity to dumps and heaps of domestic wastes put them at the highest risk of contracting infectious diseases. Uncollected solid wastes are also the primary causes of blocked city drainage channels or ditches, increasing the risk of water-borne diseases and flooding during the rainy season. Solid waste can also enter surface and subsurface waterways and pollute water resources.

One of the most disgusting and unsightly yet ubiquitous solid wastes in both urban and rural areas is discarded low-density polyethylene (LDPE) plastic bags. The quantity of LDPE plastic bags in solid wastes has increased dramatically during the past 10 years and is expected to increase as people find them quite convenient for bagging groceries and for use as trash bags. Shops usually hand out to shoppers these thin and flimsy shopping bags free-of-charge for carrying purchased stuffs. Although they do not contribute much volume to municipal solid waste, most discarded LDPE plastic bags end up in streets, steams, rivers, ditches, gardens, and parks and on fences, poles, rooftops, and trees, and, therefore pose a serious environmental hazard as well as compromise the aesthetics of the urban landscape. LDPE plastics are usually made from non-biodegradable substances that do not breakdown easily in the environment or take several thousand years to degrade. Currently, they are not recycled in Addis Ababa. Many urban households routinely throw away discarded plastic bags as litter—quite often with household waste contained in them—in street corners and ditches. Domestic animals that consume them suffer. Some urban households with no access to proper sanitation facilities, for instance Harar's old walled quarter, use plastic bags for defecation. Such waste usually ends up in street corners and thus endangers the health of urban residents, especially of children, because the waste creates conditions favorable for the survival and growth of microbial pathogens.

The use and haphazard disposal of plastic bags and its impact on human and animal health should be a matter of public concern. Many urban residents and food establishments use plastic bags not only for shopping but also for storage of cooked food. The pigments of the colored plastic bags contain toxic metals like copper, lead, chromium, cobalt, selenium, and cadmium. Thus, use of plastic bags for storage of cooked food poses a serious health risk. Some local environmentalists wish to ban the use, sale, and production of LDPE and other plastic bags and advocate the production of bags made from cheap cloths to replace them.[6]

**Disposal of Solid Waste**

Waste disposal means the depositing, dumping, injecting, or placing of any kind of waste or hazardous waste into or onto the land. From an environmental health point of view, it is extremely crucial that the site selected to dispose of any kind of waste be isolated from built environments; in addition, it cannot permit waste to contaminate surface and ground waters. There are different ways of disposing of municipal and hazardous wastes (Moeller, 1997: 183-84). *Landfills* are disposal facilities in which the waste is dumped into or onto the land. Landfills must be lined to prevent leakage and they must have structures to collect any leachate or surface runoff. *Surface impoundments* are artificial or natural basins that can be used to dispose of hazardous waste. *Underground injection wells* are concrete- and steel-encased shafts placed deep in the earth into which wastes are injected under pressure. Such facilities are commonly used for the disposal of hazardous liquid wastes. *Land treatment* is a disposal process in which sludge from municipal sewage treatment plants is incorporated into the soil surface so that naturally occurring microbes in the soil break down the hazardous components. *Incineration* is the combustion of organic waste matter, partly by the heat released as the waste burns. The residue is disposed of or used as a cover layer in landfills.

On-land dumping is the only means of solid waste disposal in the city of Addis Ababa. This is by far the least expensive means of solid waste disposal. What the city is able to collect is disposed of at Repi municipal dumpsite located in the southwest of the city, some 15 kilometers from the center. The site is located on a 2-hectare prime farmland. It was commissioned more than 35 years ago. The only factors that were considered in selecting this site were hauling distance and the availability of land. There was no evaluation of the underlying soil structure and topography, climatological conditions, surface water hydrology, or the geologic and hydro-geologic conditions of the area. The waste placed at the dumpsite each day is spread out, leveled by a bulldozer, and compacted by a steel studded wheel compactor to reduce the volume (Region 14 Administrative Health Bureau, 1997: 38). The condition of this waste disposal site is horrible. The waste is untreated. Both domestic and industrial wastes are dumped at the site. The waste is not covered with soil to prevent the harboring of disease vectors (such as flies, mosquitoes, and rats), odors, air pollution, and other hazards. Access to the dumpsite is not restricted to prevent illegal activities or dumping. The site is not protected by levees and ditches to prevent storm water flooding. It has no structures in place to collect and control runoff; therefore, nearby surface and ground waters are vulnerable to contamination by leaching. Leachate (seepage of liquid through the waste), high in biochemical oxygen demand, chloride, organics, metals, nitrates, and other contaminants, has little difficulty reaching surface and underground water courses found around the site. There is, obviously, a definite need for research on the possible effects of leachate on the surrounding water sources on which many people depend for drinking water.

Not all collected wastes find their way to the disposal site. Unable to obtain formal employment, many of the urban poor earn a living from recovering and recycling thrown away materials in Addis Ababa. Wherever city waste containers are located and at the municipal dumpsite, one sees many people, especially children, rummaging through mountains of garbage to pick out recyclable materials. Between 300-500 children and adults—derogatorily called *yeqoshe lijoch* or children of the dumps—scavenge the municipal dumpsite every day (Malifu, 2001: 148).[7] The most popular items scavenged include paper, plastics, bottles, cans, rubber, fabrics, scraps of cardboard and metals and the soles of worn-out shoes. Most of these items are sold to petty recyclers or to people who in turn sell to small industries that use this waste as raw materials. Some organic wastes are also scavenged. Many street children often find their next meal by picking through scraps dumped by restaurants and hotels. Some also collect such food scraps and rotten fruits to sell to poor people in the city. Despite the environmentally beneficial aspect of waste recycling, scavenging has an adverse impact on the health of the scavengers. Much of the city's industrial solid wastes and wastes generated from hospitals, health centers, clinics, and medical laboratories are often disposed with the regular non-hazardous and non-infectious wastes and, as a result, scavengers may sustain injuries and come into direct contact with toxic or infectious items. Contaminated scavenged articles may also come in contact with the general public (for example, reuse of waste paper to wrap food).

There is no doubt that the Addis Ababa city dumpsite is improperly managed. It attracts all types of disease-spreading insects. The stench can be felt kilometers away from the site. When the site was chosen over three decades ago, it was located at a safe distance from human settlements. Today, the site is no longer at the periphery of the city. Built areas have gradually extended toward the site and many residential neighborhoods now surround it. The communities around the dumpsite bitterly complain about the increasing odors, blowing litters, flies, rats, and pollutants spewed out by the waste. Wind-blown debris and odor end up in residential compounds where children play and adults perform household activities. No one knows the precise environmental impact of the dumpsite, but many nearby residents complain of throat and eye inflammations. In October 2000 the surrounding communities brought their complaint to the city hall and were told that the site would soon be abandoned in favor of a new dumpsite elsewhere. City sanitation engineers admit that the dumpsite has already exhausted its maximum absorption capacity and recommend its abandonment: the site poses a health risk to nearby residents through leaching to groundwater, run-off to surface water, air contamination from burning, direct contact at the disposal site, and indirect exposure through food chains.

As the city's population and consumption grow, the quantity of municipal solid waste will increase. The rate of solid waste generation already far exceeds the rate of collection and disposal. Thus, indiscriminate dumping will continue. The fact that most domestic waste is organic and prone to putrefaction and bad odors does not help matters, for people cannot tolerate the decay: rather than wait for collection, they dispose of waste in open areas.

## Healthcare Waste

Hospitals, clinics, health research institutions and laboratories generate a variety of wastes. Seventy-five to ninety percent of these wastes are considered general waste that may not pose health risk to healthcare workers or to the public. These may include paper, food, and wastes resulting from cleaning and housekeeping of the establishment. The remaining 10 to 25 percent of the wastes may pose health risks (Table 4.3). These may include infectious waste that contains pathogens (bacteria, viruses, parasites, or fungi), pathological waste (consisting of tissues, organs, body parts, human fetuses, blood, and body fluids), sharps (such as hypodermic needles, scalpels, knives, infusion sets, saws, broken glasses, and nails), pharmaceutical waste (such as expired, unused, spilt, and contaminated pharmaceutical products, drugs, and vaccines), genotoxic waste (such as vomit, urine, or feces from patients treated with cytostatic drugs, chemicals, and radioactive material), and chemical waste (consisting of discarded solid, liquid, and gaseous chemicals (WHO, 1999a: 2-5).

## Health Impacts of Healthcare Waste

Hazardous wastes generated by hospitals and other healthcare establishments can transmit diseases or cause injury. People working in healthcare establishments and associated institutions are the most susceptible to such dangers; however, poor management can expose the public at large to risks. According to the World Health Organization, those with the highest exposure to healthcare hazardous waste include: medical doctors, nurses, healthcare auxiliaries, and hospital maintenance personnel; patients in healthcare establishments or receiving home care; visitors to healthcare establishments; workers in support services allied to healthcare establishments, such as laundries, waste handling, and transportation; and workers in waste disposal facilities such as landfill or incinerators, including scavengers (WHO, 1999a: 20).

Infectious waste and sharps, for instance, can transmit pathogenic microorganism through a puncture, abrasion, or cut in the skin, through the mucous membranes, or by inhalation and ingestion. Table 4.4 shows a list of infections caused by exposure to healthcare wastes. Table 4.4 lists examples of infections that can be caused by exposure to healthcare wastes.

**Table 4.3 Categories of Healthcare Waste**

| Waste Category | Description and examples |
|---|---|
| Infectious waste | Wastes suspected to contain pathogens<br>e.g. laboratory cultures; wastes from isolation wards; tissues (swabs), materials, or equipment that have been in contact with infected patients; excreta |
| Pathological waste | Human tissues or fluids e.g. body parts; blood and other body fluids; fetuses |
| Sharps | Sharp waste e.g. needles; infusion sets; scalpels; knives; broken glass |
| Pharmaceutical waste | Waste containing pharmaceuticals<br>e.g. pharmaceuticals that are expired or no longer needed; items contaminated by or containing pharmaceuticals (bottles, boxes) |
| Genotoxic waste | Waste containing substances with genotoxic properties e.g. Waste Containing Cytostatic drugs (often used in cancer therapy); genotoxic chemicals. |
| Chemical waste | Waste containing chemical substances<br>e.g. laboratory reagents, film developer; disinfectants that are expired or no longer needed; solvents |
| Wastes with high content of heavy metals | Batteries; broken thermometers; blood pressure gauges, etc. |
| Pressurized containers | Gas cylinders; gas cartridges; aerosol cans |
| Radioactive waste | Waste containing radioactive substances<br>e.g. unused liquid from radiotherapy or laboratory research; contaminated glassware, packages or absorbent paper; urine and excreta from patients treated or tested with unsealed radio nuclides; sealed sources |

Source: World Health Organization (1999), *Safe Management of Wastes from Healthcare Activities*, Geneva: WHO, p. 3.

**Table 4.4 Examples of Infections by Exposure to Healthcare Wastes, Causative Organisms and Transmission Vehicles**

| Type of infection | Examples of causative organisms | Transmission vehicles |
| --- | --- | --- |
| Gastroentric infections | Enterobacteria, e.g. Salmonella, Shingella spp., Vibrio cholerae; helminths | Feces and/or vomit |
| Respiratory infections | Mycobacterium tuberculosis; measles virus; Streptococcus pneumonia | Inhaled secretions; saliva |
| Ocular infection | Herpes virus | Eye secretions |
| Genital infections | Neisseria gonorrhoeae; herpes virus | Genital secretions |
| Skin infections | Streptococcus spp | Pus |
| Anthrax | Bacillus anthracis | Skin secretions |
| Meningitis | Neisseria meningitides | Cerebrospinal fluid |
| Acquired Immuno deficiency Syndrome (AIDS) | Human immunodeficiency virus (HIV) | Blood, sexual secretions |
| Haemorrhagic fevers | Junin, Lassa, Ebolla, and Marburg viruses | All blood products and secretions |
| Septicaemia | Staphylococcus spp | Blood |
| Bacteraemia | Coagulase-negative Staphylococcus Spp., Staphylococcus aureus; Enterobacter, Enterococcus, Klebsiella, and Streptococcus spp. | Blood |
| Candidaemia | Candida albicans | Blood |
| Viral hepatitis A | Hepatitis A virus | Feces |
| Viral hepatitis B and C | Hepatitis B and C | Blood and body fluid |

Source: World Health Organization (1999), *Safe Management of Wastes from Healthcare Activities*, Geneva: WHO, p. 21.

A survey of 6 hospitals and 12 health centers in Addis Ababa in 2001 found that healthcare wastes are managed in a crude, haphazard manner. The survey found that:

- None of the healthcare establishments kept records of the quantities of wastes they generate.
- All of the establishments drained hazardous and non-hazardous liquid wastes untreated into the municipal sewer, septic tanks, or ditches.
- Only six establishments (Yekatit 12, Menelik, and Zewditu hospitals and Lideta, Wereda 25, and Kaliti health centers) segregated their solid wastes. Hazardous chemicals are mixed with non-hazardous solid wastes and are put away in municipal containers or dumpsites.
- None (except Gandhi Memorial Hospital) instructed staff workers, including maintenance personnel and waste handlers, how to manage healthcare wastes.
- Only two (Wereda 23 and Kaliti health centers) had a manual or guideline on safe collecting, handling, and storing of wastes.
- All establishments produced infectious wastes (lab cultures, tissues, and excreta), pathological wastes (body parts, blood, and body fluids), sharps (needles, knives, blades, and broken glasses), expired pharmaceuticals, chemical wastes (laboratory reagents, expired disinfectants, and solvents), and general non-hazard wastes (i.e., wastes from administrative offices and maintenance of establishment premise).
- On the positive side, staff workers handling wastes used gloves in all establishments but one.
- Eleven establishments incinerated some of their hazardous wastes, including discarded syringe needles, swabs, bandages, plastics and other types of infectious wastes. Three establishments incinerated all of their wastes. Incineration is considered to be the best method of disposing of hospital wastes, because it kills pathogens and reduces the volume of waste for ensuing burying or disposal at a safe place. However, nearly all the establishments acknowledged to have disposed of hazardous wastes into municipal dumpsters along with non-infectious wastes, especially when incinerators malfunction or exceed their capacity to handle all hazardous wastes generated on the premise. The establishments that do not segregate their healthcare wastes are especially guilty of such unsafe practices as disposing hazardous wastes in a municipal dumpsite. Such slapdash practices place people who scavenge the municipal dumpsite for their livelihood in direct contact with all types of dangerous healthcare wastes. Residues from incineration processes—including the non-combustible portion of the waste—are dumped in municipal containers.
- Four establishments practiced open burning of wastes on their premises. Such burning often included dangerous wastes such as sharps that may not be destroyed even at high temperatures.

- Half of the establishments buried pathological wastes (such as body parts from surgical operations, blood, placenta, and fetuses) inside their compounds.
- Only one establishment (Wereda 23 Health Center) was aware of legislation applicable to healthcare waste management.

The survey indicates the absence of safe waste management in the city's healthcare establishments. A national law on the management of healthcare waste does not exist. Healthcare institutions are not obliged to keep records of their generation and disposal of wastes. Healthcare institutions are not inspected to ensure safe handling, storage, and disposal of healthcare wastes. The Environmental Protection Authority has produced policy documents outlining the need for laws on healthcare waste management; nevertheless, the federal government and its legislatures have made no motion to introduce new laws governing the safe management of healthcare wastes and the protection of health personnel's health and safety. The Federal Government's Public Health Proclamation No. 200/2000 states: 'Any solid, liquid and other wastes generated from hospitals should be handled with special care and their disposal procedures should meet the standards set by the public health authorities'. However, none of the health institutions surveyed for this study acknowledged the existence of a document outlining healthcare waste management policy and procedures issued by public health authorities.

## Improving Solid Waste Management

It is clear that all Ethiopian urban centers lack adequate solid waste collection and disposal services. The problem is a serious concern because it threatens public health and the living and working environment. In 1996, the Hygiene and Environmental Department of the MOH made recommendations to ameliorate the problem (MOH, 1996: 77-86). The recommendations were too broad, but they point in the right direction. The recommendations underscored:

*The need for institutional arrangement for planning, management, and service delivery* While the Municipalities should remain the sole responsible government organization for solid waste management, their functions should be managed by one centralized agency that has an equal status with other municipal agencies and units. A special unit should have the authority to assess and justify finance, equipment, and human power needs. The Municipalities currently plan and administrate social, environmental, health, and educational activities. However, solid waste management services in the private sector ought to share these responsibilities. The involvement of the private sector and the introduction of skilled staff into management would improve the level of service rendered.

*The need for financial resources and budget planning*  Most municipalities do not have financial budgets for regular renewal and expansion of their collection fleets. Finances to the solid waste program are allotted primarily in times of crisis. Financial arrangements should ensure a reliable source of money for regular operation and maintenance of equipment. This source could be from the municipal budget or from user fees. User fees, along with electricity and water fees, are recommended, first, because joint billing costs less, and second, because it is easier to terminate water and electricity services for non-payment than it is to terminate solid waste disposal services. Direct user charge revenue is not recommended, as it is difficult to enforce.

*The need for selecting appropriate technology*  Appropriate technology is an important consideration in efficient solid waste management. Selection should be based on urban setting, types of solid waste needed to be disposed of, and land available for disposal. Issues of worker health and safety and limitations on foreign currency should also be considered important factors.

*The need for enforcement support systems*  Community participation is important. Community cooperation can be gained by establishing appropriate legislation. Laws and regulations coupled with adequate service and enforcement are needed to ensure cooperation of residents. Rules on solid waste management have been in existence since 1940 and can still be used along with additional standards, including the enactment of national or regional regulatory standards, to protect water sources from pollution. Indiscriminate dumping of solid waste is a serious threat to water quality and depreciates the recreational value of the region. Waste reduction at the source should be encouraged through reuse and recycling activities, as they result in savings to the city/town.

*The need for cooperation between institutions*  Cooperation between the various organizations vested with power and duty in the matters of solid waste disposal is extremely important to the promotion of an efficient system. This is particularly true among the EPA, the MOH and the ESA. The EPA is charged with the preparation and implementation of standards regarding environmental policy and the protection of soil, water, air and the biological systems they support; the recommendation for the application of regulations for better environmental protection; and provision of instruction to enhance awareness of the need for environmental protection. The MOH is responsible for quarantine controls to protect public health. The ESA is also responsible for setting standards and the Municipal administrations are directly involved in the management of solid waste.

*The need for providing public education and participation program*  An education program to improve people's understanding of the necessity of sanitary solid waste disposal is imperative. The program should emphasize the importance of a clean environment and the health, social, and economic risks of uncontrolled solid waste disposal. Education is possible through pamphlets, newspapers, radio, and other such media. Sanitation campaigns are also valuable for specific emergency events.

*The need for establishing incentive and disincentive for workers*    Incentive and disincentive systems for workers can improve the overall efficiency of work. Rewards and responsibility should be provided to workers so that they have a larger and more active role in ensuring good work ethics. Important incentives include a clear designation of responsibility and authority. All individuals/crews should be given specific responsibilities and their achievements should be monitored and evaluated.

*The need for monitoring and evaluation*    Monitoring and evaluating the system is essential for the successful working of the solid waste management program and the attainment of its goals. The monitoring system should show coverage—served and unserved populations, costs and use of technologies. The success or failure monitored can then be promptly corrected, each step of the way.

## Youth, Music, and the Environment

In recent years an environmental awareness movement has emerged in Addis Ababa because of the municipal authority's inability to carry out the tasks of maintaining a clean, healthy urban environment. This movement began by a middle-aged poet, storyteller, musician, and composer named Seleshi Demissie. Upon his return from a twenty-three year exile abroad, he was appalled by what he saw in the capital city of his nation and decided to do something to improve the environment. In the late 1990s, he launched an environmental youth movement in Addis Ababa.[8]

Seleshi first created a character called Gashe Abera Molla through a musical drama that contained messages about health, sanitation, and culture. Seleshi Demissie assumed the character of Gashe Abera Molla and began to entertain the social life of the urban community as a musician, a poet, and a philosopher. People came to know Seleshi as Gashe Abera Molla, and his character became famous throughout the city. The experience offered Seleshi the opportunity to meet and talk to many people across various communities in Addis Ababa. All the conversations he had revolved around what could be done to improve the sanitation problems in the city. After he received some feedbacks he designed a project called *Youth, Music, and the Environment*. He not only thought the youth could be a very good target group to convince, but he also observed the vast majority of them wandering around the city, idling in neighborhood cafes, and chewing *qat* in street corners or playing soccer in open spaces littered with human excrement and household garbage. Most of these young people were students, but many were without jobs. He saw the potential of the youth to reverse the deterioration of the city. He felt that with motivation and organization the youth could work wonders. In an interview I conducted in January 2003, Seleshi described, in his own words, the genesis and accomplishments of his environmental movement as follows:

'I felt that the youth could relate their life to the environment; they could use the environment for their own benefit. They could keep the environment clean and gain health benefits. They could learn from their experience and even make a living by participating in environmental causes or projects. I wrote a letter to the minister of education to allow me to teach the youth and to draw them to the *Youth, Music, and the Environment* project. The Ministry of Education was happy about my idea and became supportive. I recruited 18,000 students from 26 high schools in Addis Ababa. Then I had to find a large place where I could gather them for learning sessions. Halls at the National Theater, Hager Fikir, the University and some high-school auditoriums were accessed. I finally took about 13,000 of the 18,000 students (500 students from each of the 26 schools); I could not accommodate all of the 18,000 students because of lack of space. I divided them into groups and ran sessions from Monday to Sunday. The group lessons involved music, literature and drama, dance (traditional and modern), poetry, and painting. As the lessons went on, I interjected the idea of good health. I said to them that to do all that they like in life, they need good food, clean air, and a clean environment. If they don't have good health, they cannot be a good musician, a poet, a dramatist, a dancer, or a painter; they cannot be anything. I told them that the environment in which they live affects their life; if the environment is healthy, they will be healthy. They get clean water and clean air from a clean environment. My message underscored the importance of an unpolluted environment and the role of solid waste in the transmission of diseases such as cholera, typhoid, malaria, etc. The youth never realized that the environment is the most important thing for a healthy life. To them the environment has nothing to do with their lives. The environment is the last thing they think about in life. I wanted to change this apathetic mentality. I wanted to create consciousness among the youth about their environment. They eventually started to get the idea. We then started to discuss how to reduce waste, how to manage waste, and how to recycle some throw-away things. After a year of education and consciousness-building these kids went back to their communities and started to spread this idea. Many started spearheading environmental clean-ups in their communities. The youth couldn't believe that every one in the city lived in fouled environments. Even the so-called rich communities in Bole area and elsewhere in the city live in unhygienic environments. They drive nice cars but they live in a polluted environment. In rich neighborhoods in the Bole area we found piles of garbage: diapers, bones, cans, vegetable and fruit peels, festal, napkins, and broken household stuffs. These rich people live in these huge, wonderful houses but they don't care about the stinking rubbish around them.

In the last year or so, we selected the worst polluted areas as pilot clean-up projects, like Tekle Haimanot areas and the Ras Mekonen Bridge. The latter was perhaps the filthiest bridge to cross in the city. The unpleasantness of the sight and the foul odor disgorged from around the bridge more than offended the senses. The spot had been a public health threat for years. The municipality did nothing, even when dining establishments that are located right by the bridge were dangerously exposed to hordes of flies and other disease vectors that consumed the feces and household waste thrown in and around the river. People had to look the other way and clamp their mouth and nose with handkerchiefs or shawls as they crossed the bridge. We cleaned all that filth. It took us five months.

Cleaning these places was not fun at all. It was a hellish job. The stench was sickening and nauseating; it was unhealthy. But somebody had to do it. The result of these clean-up projects amazed everybody in the city, including those who did not take us seriously in the first place.

Our next plan was to go beyond the youth and draw the rest of the citizens, the government, the business sectors, and indigenous non-governmental organizations into this environmental movement. Initially, government leaders, especially municipal leaders, were scared about the movement; they felt I was going to overthrow them. They were quite nervous and jealous at the same time. After they saw all that the movement had accomplished thus far, they began coming to us. They felt that this environmental movement was a big fire. They have apparently decided not to mess with this big fire and, instead, flow with it. All those officials who were belittling the movement have now turned around to become friendly. They have recently expressed their willingness to support and work with the movement. At this point in time, our environmental movement has gone beyond Addis Ababa and is reaching other major cities like Harar, Dire Dawa, Jimma, Jijiga, Arba Minch, and Awasa. I intend to make it a national movement.

At present all our pilot projects are blooming and the government is getting involved in some ways. Now we have developed an action program called 'Clean and Green Addis Initiative' that involves the participation of local NGOs, the government, private enterprises, and communities. We approach organizations—large and small—to sponsor clean-up projects in the city. At the present time, the Ethiopian Airlines, Unity College, and Kiray Betoch are sponsoring the Bole Road clean-up project. These organizations not only cover the cost of the clean up, but will also be responsible for up-keep in the future. We spearhead and coordinate this and other projects. We encourage communities to get involved in cleaning nearby parks, open spaces, green areas, and ditches.'

Our movement does not have many financial resources. We recycle things to raise some funds. We also get some funds from embassies. We do not receive any public funding. We do not receive funding from corporations either. We do not solicit financial support from international NGOs. We do not believe we should. The number of foreign NGOs working in our country has increased radically in the last few years, but not many visible things have been achieved as a result. NGOs have been involved in some development activities, health and water, but those activities have not made the lives of people any better. They focus too much on emergencies and contribute very little to long-term development efforts. They foster dependency, not self-sufficient development. They don't empower people and always want to be the boss. Our movement wants to fight against the NGO-dependency mentality. We want to involve local NGOs, even though at the present time many are not ready to get involved because they have their own narrowly focused individual agendas to pursue.

The success of this project is the awareness it has created in the youth. One cannot change this country for good in two or three years. It will take a long, long time and it will require the participation of each and every citizen in this effort. What this movement has done thus far is immensely great. The youth believed in the movement and in themselves. We have made some significant changes and even the government is now convinced that we can make a difference in improving the city's environment. The municipal people now want to work with us. But there is some misconception about what I'm doing. Some people think that I'm a millionaire, and they want to see me do everything while they sit and watch. I've some frustration in this regard. We have lots of work left to do, especially in raising environmental awareness among the general public. Most urban residents lack understanding of the connection between the health of their environment and their own health.'

It is evident that Gashe Abera Molla's environmental campaign in Addis Ababa has been a big success. In many neighborhoods, clean-up programs have been set up on a regular basis. Drainage channels have been cleaned in several places. Many public places that had been littered with feces and rotting household wastes have been cleaned up and are now bedecked with beautiful plants and flowers. In partnership with *Lem Ethiopia* (another local non-governmental organization involved in reforestation projects), his movement has planted over 200,000 trees in hilly areas around the city to prevent soil erosion and flooding. Thanks to his movement, six more parks have been added to the city. He has also promoted recycling. For instance, truck tires that would normally be thrown away or burned are recycled and transformed into flower pots. His latest innovation is the introduction of mobile toilets that can be accessed for a fee of 0.10 Birr per visit. The overall goals of the toilet project include: lessening human waste pollution and its effects on the health of the people, by employing street children of both genders to manage these toilets, with the hope of eventually upgrading to include full showers, and turning over the ownership to the youth, thereby improving their quality of life; the start up of small waste collection businesses that will create more jobs; and motivating government and other local NGOs to become involved in creating a better, healthier, cleaner, and more civically responsible city for future generations.[9]

Despite the effort by the Gashe Abera Molla movement, many environmental problems remain unsolved and require action beyond the efforts of this movement. Neighborhood and street clean-up programs solve only a few of the scores of environmental problems the city faces. They cannot improve the city's housing conditions or human waste problems. These problems have considerable impacts on the health problems of the city's residents. But still, the achievements of Gashe Abera Molla's environmental movement have been remarkable. The movement has been able to raise consciousness and awareness of urban environmental issues among the urban youth. Some criticize the movement for not coordinating its plans and activities with urban municipalities. But Gashe Abera Molla should not be faulted for this because the bureaucrats in most municipal halls failed to appreciate his genuine effort. The Addis Ababa municipal officials did everything they could to belittle or undermine his movement, for they not only coveted his accomplishments, but he also exposed their utter incompetence to manage the city's sanitation problems. The current one-party municipal government leadership is antagonistic to the emergence of autonomous civil society organizations for fear that such organizations may turn the public against them.

The Gashe Abera Molla environmental movement has been able to cultivate youth followers, but it has yet to succeeded in drawing the support and participation of the general public. Most urban residents seem to appreciate what he is doing, but few have acted upon his call or joined his movement. Additionally, the movement has yet to solicit the support and participation of a wide range of institutions of civil society such as other environmental groups, private businesses, professional and trade associations, and community organizations. The movement also needs to encourage the establishment of neighborhood-based citizen groups that can take responsibility for periodic neighborhood clean-ups and for sustaining rehabilitated public spaces. Without continued education and awareness programs and the participations of neighborhood residents and business owners, the city could quickly return to its previous conditions.

## Solid Waste Minimization Through Composting

Solid waste disposal problems in urban areas will only increase, not diminish. As personal incomes increase, urban households will consume far more resources (food, water, energy, construction materials, and communication devices). As a result, far more wastes will be generated. It is, therefore, essential to search for ways by which the solid waste load in the city can be reduced. In this regard, composting can be an effective means for alleviating the problems of unmanaged waste in the city. Each urban household generates and throws away large amounts of solid wastes that contain organic matter. In the case of Addis Ababa, about 60 percent of the solid waste is organic matter (Addis Ababa City Government, 2001: 58). It is this organic part of the waste that decomposes more quickly and attracts pathogens that cause disease. Household organic wastes can be considered to contain proteins, amino acids, carbohydrates, cellulose, lipids, and ash. Subjecting these organic materials to aerobic micro-bacterial decomposition produces a humus material commonly known as compost (Tchobanoglous, et. al., 1993: 302). Many rural farmers process household wastes for use on their farms. This practice, however, has not caught on in urban areas where household wastes are fast growing. The processing of compost in urban areas can take place only if wastes are sorted out. It seems that the most effective method of segregating or sorting out wastes is at the household level.

Composting has many economic and environmental benefits. First, it reduces the volume of solid wastes generated in urban areas by transforming the 'biodegradable organic materials into a biologically stable material' (Tchobanoglous, et. al., 1993: 302). Second, it improves the urban physical environment resulting from reduced waste dumping in drainage channels; thus, it lessens flooding, odors from rotting organic waste, and waste blocking streets and neighborhood roads and footpaths (Peters, 1998: 10). Third, composting destroys pathogenic bacteria, viruses, and helminthic ova since the 'biological reactions occurring during composting will convert the putrescible forms of organic waste into stable, mainly inorganic forms which would cause little further pollution effects if discharged on to land or into a water course' (Polprasert, 1996: 7). Fourth, it improves crop yield by enriching the soil with important nutrients such as nitrogen (N), phosphorus (P), and potassium (K) (Tsige, 2001). When added to soil, compost conditions the soil for long periods of time, as it decomposes slowly. The sustained activity of the compost increases the water content and retention of the soil, increases soil aggregation, aeration, permeability, and water infiltration, and decreases surface crusting (Polprasert, 1996: 110). Fifth, it helps reduce the detrimental effects of extreme acidity or alkalinity, or the use of chemical fertilizer. Thus, the benefits of compost processing are not limited to mitigating the waste problems in urban areas. If the composting operation is done properly, the product can be used as a fertilizer on household garden plots or, where such activities are constrained by the lack of space, it can be sold to urban and peri-urban farmers (Ketema, 2000). In this regard, composting activities have the potential to generate income for some creative urban entrepreneurs. Although such activities will not solve the urban waste management problems, they will make a difference.

**Sanitation and Health**

The sanitation situation in urban Ethiopia is deplorable. More than 90 percent of the excreta disposal is via dry latrines and open fields. The vast majority of the population in small- and medium-size towns defecates in streams, drains, or open fields. Piped sewerage systems are virtually non-existent in urban areas. Even the capital has an inefficient system to which only a small percentage of the population is linked. The only other town with some semblance of sewerage system is Gondar. Flush toilets, septic tanks, and soakage pits are a rarity. Flush toilets are usually found in large town hotels, government offices, and a few high-income residential houses. Liquid wastes from all urban sources are discharged untreated into ditches and nearby waterways. Well over half of the solid wastes generated in cities and towns are disposed of haphazardly in street corners, open spaces, and streams.

Table 4.5 shows the diseases that are prevalent in poor sanitation conditions and how they are transmitted. Viruses, bacteria, protozoa, and helminthes (worms) cause these excreta-related diseases. In his *Low-Cost Urban Sanitation*, Duncan Mara (1996: 12) classifies these diseases by grouping them according to their common environmental transmission routes. According to him, successful transmission of an excreted infection depends on the number of pathogens excreted. The changes in the number of pathogens during environmental transmission depend on specific properties of the pathogen such as: *Latency* or how long it takes for the pathogen, once excreted, to become infective; *persistence*, or how long the pathogen can survive in the environment;

*multiplication*, the ability of the pathogen to increase its numbers whilst in the environment; *infectivity*, which is the probability of one pathogen initiating infection;' and *host susceptibility*, or the degree of susceptibility to the disease.

Mara's environmental classification of excreta-related diseases provides seven categories (Table 4.5). Category I includes non-bacterial fecal-oral diseases: all the excreted viral and protozoa diseases and two helminthic diseases. The transmission features of these excreted pathogens are: non-latent, low to medium persistence, unable to multiply, highly infective, and no intermediate host.

In Category II are bacterial fecal-oral diseases and their transmission features are described as non-latent, medium to high persistence, able to multiply, medium to low infectivity, and no intermediate host.

Category III includes Geohelminthiases or soil transmitted nematode worms (*Ascaris lumbrcoides*, the human roundworm; *Trichuris trichiura*, the human whipworm; *Ancylostoma duodenale* and *Necator amercanus*, the human hookworms; and *Strongyloides stercoralis*, the small nematode. Their transmission features are characterized as latent, very persistent, unable to multiply, very high infectivity, and no intermediate host.

Category IV, Taeniases, contains the two principal human cestode worms, including *Taenia saginata* (the beef tapeworm) and *T. solium* (the pork tapeworm). The transmission features of these pathogens are characterized as latent, persistent, able to multiply, very high infectivity, and cow or pig intermediate host.

Category V includes water-based helminthiases (all the water-based human trematode worms) of which the major ones are *Schistosomiasis mansoni*, *S. japonicum* and S. *haematobium* (the main human *schistosomes* or blood fluke). Their transmission features are described as latent, persistent, able to multiply, high infectivity, and intermediate aquatic hosts (snail, fish).

Category VI includes excreta-related insect-vector diseases—flies, cockroaches and other insects facilitate the transmission of Category I-III infections by carrying viruses, bacteria, and nematode eggs on their bodies or in their intestinal tracts, from feces to food.

Category VII includes excreta-related rodent-vector diseases that contain all Category I-III infections the transmission of which can be facilitated by rodents (rats, for example) that carry the excreted pathogens in their intestinal tracts or their bodies from feces to food or drink (Mara, 1996: 17-27).

**Table 4.5 Environmental Classification of Excreta-related Diseases**

| Category | Environmental transmission features | Environmental transmission features | Environmental transmission focus |
|---|---|---|---|
| I. Non-bacterial fecal-oral diseases | None –latent<br>Low to medium persistence<br>Unable to multiply<br>High infectivity<br>No intermediate host | *Viral:*<br>  Hepatitis A and E<br>  Rotavirus diarrhoea<br>*Protozoan:*<br>  Amoebiasis<br>  Crystosporidiasis<br>  Giardiasis<br>*Helminthic:*<br>  Enterobiasis<br>  Hymenolepiasis | Personal<br>Domestic |
| II. Bacterial fecal-oral disease | None –latent<br>Medium to high persistence<br>Able to multiply<br>Medium to low infectivity<br>No intermediate host | Campylobacteriosis<br>Cholera<br>Pathogenic *Escherichia coli* infection<br>Salmonellosis<br>Shigellosis<br>Typhoid<br>Yersiniosis | Personal<br>Domestic<br>Water<br>  Crops |
| III. Geohelminthiases | Latent<br>Very persistent<br>Unable to multiply<br>No intermediate host<br>Very high infectivity | Ascariasis<br>Hookworm infection<br>Strongyloidiasis<br>Trichuriasis | Peri-domestic<br>Field<br>Crops |
| IV. Taeniases | Latent<br>Persistent<br>Able to multiply<br>Very high infectivity<br>Cow or pig intermediate host | Taeniasis | Peri-domestic<br>Field<br>Fodder crops |
| V. Water-based helminthiases | Latent<br>Persistent<br>Able to multiply<br>High infectivity<br>Intermediate aquatic host(s) | Schistosomiasis<br>Clonorchiasis<br>Fasciolopsiasis | Water<br>Fish<br>Aquatic species or aquatic<br>Vegetables |

| VI. Excreta-related Insect-vector diseases | Infections in I-III transmitted mechanically by flies and cock-roaches | Peri-domestic |
| | Bancroftian filariasis transmitted by *Culex quinquefasciatus* | Water |
| VII. Excreta-related rodent-vector diseases | Infections in I – III Transmitted mechanically by rodents | Peri-domestic |
| | | Water |
| | Leptspirosis | |

Source: Duncan Mara (1996), *Low-Cost Urban Sanitation*, New York: Wiley, pp. 18-19.

The control methods of all these categories of infections are listed in Table 4.6. It is quite apparent that improving water supply along with sanitation affords the most beneficial improvements in health. Most of the important excreta-related diseases can be prevented or significantly reduced with the provision of water that is of sufficient quantity and quality, provided people practice proper hygiene. Studies suggest that the effect of hygiene practice on water- and sanitation-related diseases can be as enormous as that of improvements in water supply and sanitation. Hygiene behavior is a crucial influence on the transmission of diseases at various phases. 'A primary barrier is safe defecation, to prevent fecal pathogens from entering the human environment. A secondary barrier is hand washing, to ensure that fecal contamination on hands is not transmitted via food' (Adams, 1999: 99). Clearly, improvements in water and sanitation can synergistically yield larger health impacts than either alone (Esrey, 1996: 618).

**Table 4.6 Control of Excreta-related Diseases**

| Category | Major control interventions |
| --- | --- |
| I. Non-bacterial fecal-oral   diseases | Improved water supplies<br>Hygiene education<br>Improved housing<br>Provision of toilets |
| II. Bacterial fecal-oral diseases | Same as above, plus:<br>Treatment of excreta or sewage prior to discharge of reuse |
| III. Geohelminthiases | Provision of toilets<br>Treatment of excreta or sewage prior to land application |
| IV. Taeniases | As for III, plus:<br>Meat inspection<br>Through cooking of meat |
| V. Water-based helminthiases | Provision of toilet<br>Treatment of excreta or sewage prior to discharge or reuse<br>Control of animal reservoirs<br>Snail control<br>Reduce water contact |
| VI. Excreta-related insect-vector disease | Domestic and peri-domestic hygiene<br>Insect control |
| VII. Excreta-related rodent-vector disease | Peri-domestic hygiene<br>Trapping<br>Rodenticide application |

Source: Duncan Mara (1996), *Low-Cost Urban Sanitation*, New York: Wiley, pp. 18-19.

**Conclusion**

Poor environmental conditions pose the greatest threat to the quality of life of urban residents in Ethiopia. Pollutants are pervasive both in and out of the home. Within the home, overcrowding and inadequate water supplies and sanitation facilities provide breeding grounds for infectious diseases of all kinds. Outside the home residents' health is threatened by uncollected household solid wastes, poorly constructed and maintained pit latrines, and defecation in open spaces that provide ideal habitat for rats and flies that facilitate the fecal-oral route of many diarrheal diseases. At present, municipal governments are unable to provide adequate waste collection services and to keep the urban environment clean.

Community participation is imperative to meet the hosts of urban environmental challenges and to make urban areas livable and sustainable in the long run. Programs designed to improve the sanitation conditions in urban areas are most effective when they enlist the involvements of resident communities; collaboration allows development of innovative waste management programs that meet the needs of communities. The participation of urban communities also reduces the cost of environmental improvement and maintenance: communities are more likely to contribute their labor or pay for maintenance costs when they notice improvements in environmental conditions. Encouraging neighborhood community organizations to participate in periodic cleaning of neighborhood streets and drainage channels and the maintenance of sanitation facilities are steps in the right direction to tackle urban sanitation problems. In addition, municipalities ought to contract out some or all of the collection of waste and sanitation services to local private enterprises. Environmental sanitation services in which local private enterprise can participate could include street sweeping, cleaning of storm drains, maintenance of parks and other public spaces, recovery of recyclables, and management of public toilets and disposal sites. Using local private organizations to deliver municipal services requires that municipal administrations become skilled in contract management to ensure quality of goods and services and prevent corruption and bribery in the contract selection.

## Notes

[1] Adequate sanitation simply refers to a facility or ordinary pit latrine, without taking into account the condition of such a facility.

[2] Cited in Ministry of Water Resource (1998a), *Comprehensive and Integrated Water Resources Management*, Volume 1, Addis Ababa: MOWR, p. 27. Unless otherwise indicated, the discussion in this section benefits from statistical information obtained from the following public documents: CSA (1995), *The 1994 Population and Housing Census Result for Addis Ababa*, Addis Ababa: CSA; AAWSA (1996), *Master Plan for the Development of Liquid Waste Facilities for the City of Addis Ababa*, Addis Ababa: AAWSA; Region 14 Administration Health Bureau (1997), *A Comprehensive Overview of Addis Ababa Municipality Solid Waste Management and its Environmental Health Inspection Services*, Addis Ababa: Region 14 Administration; MOH (1996a), *Study of Liquid Waste Management in the Urban Centers of Ethiopia*, Addis Ababa: Hygiene and Environmental Health Department.

[3] *Addis Tribune* (2000), 'Editorial: The Streets of Addis.' Available at: http://www.addistribune.com/Archives/2000/08/25-08-00/Edit.htm.

[4] The city-state of Harari has another major problem: it has no place to dump the solid wastes it generates. The city reluctantly abandoned a potential site in Hamaressa, in neighboring Oromiya, due to fierce local opposition. On the first day of shipment, the local people smashed the windshield of the garbage truck, beat up the driver and his sanitation assistant and forced them to go back to Harar with their garbage, warning them that they would slit their throats if they ever came back to the site to dump 'Harari garbage.' The warning worked: no one has ever come back to Hamaressa village with a garbage truck since. This incidence may sound like a case of NIMBY (not in my back yard) syndrome, but it is not. It stems from the old Oromo-Harari hostility that has glaringly resurfaced since the creation of ethnic-based regional states in the early 1990s. At present, the city of Harar dumps the solid wastes it manages to collect haphazardly and recklessly on Mount Hakim, a scenically beautiful mountain overlooking the city.

[5] In Dire Dawa, the second largest city in the country, people routinely dump their solid wastes into the dry riverbed that recharges the city's water supply wells without fear of legal consequences. The city collects only less than 40 percent of the solid wastes it generates.

[6] To influence the government they cite examples of African countries that have banned or are in the process of banning plastic bags. In 2003, for instance, South Africa made the use of plastic bags with a minimum thickness of 30 microns (one micro equals one-thousandth of a millimeter) illegal (BBC News, 'South Africa bans plastic bags', 9 May 2003. Available at: http://news.bbc.co.uk/2/hi/africa/3013419.stm).

[7] Besides the *yeqoshe lijoch*, there are others who are making a living by buying recyclable metal and plastic containers from households and selling them in the market for profit. Plastic containers are in high demand and are reused to hold water and other liquid needed in the home. Similarly metal aerosol containers are molded into kerosene lanterns and small containers for home use.

[8] Sileshi Demissie completed his primary and secondary education at Menelik II School in Addis Ababa. He received a diploma from the Harar Teachers Training School and taught for two years in then Wello province in the 1970s. He left Ethiopia for France via Djibouti during the revolutionary turmoil of the late 1970s. After two years of stay in France he migrated to the United States where he lived for 23 years before returning to Ethiopia in the late 1990s. While in the States, Sileshi studied ethnomusicology at Temple University and African music and musical traditions at Vermont College for four years. He says his interest in music started at home; his father played the *Krar* and he grew up listening and learning while his father played. He also had a brief music lesson at the Addis Ababa Yared Music School. As a professional musician playing the K*rar*, Sileshi has traveled to many countries in Africa, Asia, Europe and North America to perform on stages.

[9] The latest profile of Gashe Abera Molla Environmental and Development Association (GAMA) is available at:
http://www.wateraid.org.uk/other/startdownload.asp?openType=unforced&documentID=6 72.

# Chapter 5

# Pollution

As the Ethiopian economy evolves through expansion of the industrial sector, pollution poses an increasingly serious problem for both the environment and the health of the urban community. Most of the industries in the country and the pollution they produce are concentrated in Addis Ababa. With few laws that regulate environmental pollution and minimal monitoring of environmental pollution and ecosystem health, the city has little effective environmental protection. Industries have not been held accountable for their impacts on the urban natural and human environment. A common misconception within Ethiopia's industrial sector is that effluents (in liquid, gaseous, or solid forms) contain an insignificant amount of chemicals harmful to human and ecosystem health. This belief could not be farther from the truth.

Pollution is organic or inorganic matter whose presence in the environment creates a health risk. Sewage and other liquid wastes from the domestic, industrial, agricultural, and urban uses of water cause pollution when improperly controlled and disposed of. Human health and well-being can be adversely affected by environmental pollution due to contaminated air, water, or food.

## Industrial Pollution

Industrial pollutants run into the thousands. They include solvents, detergents, heavy metals, sulfides, ammonia, dyes and pigments, cyanides, mineral and organic acids, fats, salts, bleaching agents, and nitrogenous substances. Many of these pollutants have detrimental effects on the environment and either directly or indirectly harm human health when waste containing these pollutants is improperly treated, stored, transported, or disposed of.

Governments in industrial nations insist on high environmental standards: they scrutinize industrial operations. This is not the case in a developing country like Ethiopia. In most cases legal requirements are often lower or totally absent and foreign industrial investors can easily get away with any kind of environmental damage stemming from their operations. Where some kind of legal requirements do exist, foreign industrialists usually apply different standards to their operations in the developing countries from those in their home country (Larkin, 1998). In Ethiopia, foreign and domestic industries invariably release their wastes into the environment even though the law demands that their activities comply with conditions stipulated in environmental protection laws. Industries often dump their wastes haphazardly, knowing that no regulatory agency will monitor or supervise their activities (EPA/UNIDO, 2001a: 7). At present, strict implementation of the environmental laws is being sacrificed in the interest of immediate economic gains.

Addis Ababa experiences the highest health risk from industrial pollution in the country. The concentration of industries in the city has created a growing threat to health and the environment. Industries in the city operate with little or no concern for the environment. No one monitors or controls industrial pollution: liquid waste treatment plants rarely exist and where they do, they are often dysfunctional; water quality monitoring programs have not been implemented; enforcement of legislation is lacking; and regulations for planning stages and operational stages of industries are missing. In the absence of on-site treatment plants or storage facilities, industries are major polluters of the waterways in and around the city.

Ethiopia's industrial base is low, but it has grown steadily in recent years. In 1975/76, there were about 430 manufacturing industries with ten or more employees. This number was reduced to 273 with the secession of Eritrea in 1991. However, in the post-1991 liberalization and market reform the growth of industrial establishments hastened, and by 1998, there were about 741 industrial establishments employing 10 or more people with a total workforce of over 93,000 (EPA, 2001: 5). The main industries in and around Addis Ababa and other urban areas are breweries, textile mills, tanning and leather manufacturing, plastics, pulp and paper, chemicals (alum, caustic soda, soda ash, sulfuric acid, etc.), food processing, soft drink manufacturing, oil mills, dairies, slaughterhouses, metals, ceramics, pharmaceuticals, paints and dyeing, glass, batteries, and soap and detergents. Light industries (food, beverages, textiles, apparel, and leather industries) account for over two-thirds of the total industrial value added.

About 65 percent (484) of the industrial establishments, 54 percent of the gross value of industrial production, and 58 percent (54,284) of the workforce were located in Addis Ababa in 1998 (Table 5.1). The remaining industrial establishments were distributed in the regional states of Oromiya (101), SNNPR (57), Amhara (44), Tigray and Dire Dawa (21 each), Harari (6), Gambela and Somali (4), and Afar (3) (EPA/UNIDO, 2001a: 4). While Addis Ababa continues to dominate, other cities like Bahir Dar, Kombolcha, Mekele, and Awasa have emerged as industrial centers in more recent years.

Most industries need ample amounts of water and a place to discharge their liquid and solids wastes. It is thus no coincidence that most of the industries in Addis Ababa and other urban areas are located near or along waterways. Those industries that are located close to the rivers discharge their liquid waste directly into the rivers (Table 5.2). Others that are a little farther away from the rivers discharge their effluents into nearby open ditches or municipal drainage that

**Table 5.1 Distribution of Manufacturing Industries by Regional States**

| Regions | Major Product | Gross value production (000 Birr) | Number of industrial facilities | Number of Employees |
|---|---|---|---|---|
| Tigray | Textiles, medicine, marble, food | 78,193 | 21 | 1,889 |
| Afar | Cotton milling, non-metallic mineral products | 67,736 | 3 | 285 |
| Amhara | Food and beverage, textiles, non-metallic mineral products | 261,125 | 44 | 8,107 |
| Oromiya | Leather, cement, food and beverage, marble, furniture and wood products, chemicals | 1,894,373 | 101 | 15,992 |
| SNNPR | Textiles, ceramics, food, furniture and wood products | 187,833 | 57 | 5,454 |
| Gambela and Somali | Wood products, non-metallic mineral products | 3,850 | 4 | 132 |
| Harari | Food, printing, non-metallic mineral products | 91,766 | 6 | 1,000 |
| Addis Ababa | Leather, textiles, food and beverage, printing, metal products, machinery and equipment, chemicals | 3,233,762 | 484 | 54,284 |
| Dire Dawa | Textile, cement, food and beverage | 177,561 | 21 | 6,023 |
| | *Total* | 5,996,199 | 741 | 93,166 |

Source: UNIDO (2001), *Ecologically Sustainable Industrial Development Project*; US/ETH/99/068 Ethiopia, Addis Ababa: Environmental Protection Authority, p.4.

finally end up in city rivers. Still others that are located nowhere near waterways dispose their waste in nearby open space (Addis Ababa City Administration, 2000: 24). In 1993, the total amount of industrial liquid waste discharge into watercourses within the metropolitan limits of the city of Addis Ababa was estimated at 12,000 cubic meters per day, an amount equivalent to about one-tenth of the total liquid waste produced in the city (AAWSA, 1993: 79).

Industries are generally not equipped with treatment plants. In the very few cases where treatment plants are installed the facilities are usually poorly maintained, inadequate, and incorrectly operated. A 1997-1999 survey of 60 industries in Addis Ababa by the Addis Ababa Bureau of Environmental Protection found that 80 percent of the industries discharged their liquid and solid wastes into the environment. Nearly all industries produced some form of particulate or gaseous air pollutants. Only 4 industries had proper liquid waste treatment facilities (2 chemical, rubber, and plastic industries and 2 textile industries), two of which were operational at the time of the survey. Only 13 industries (21 percent) were able to quantify how much liquid or solid waste they produced (AACG, 2000: 19). For the most part, solid wastes are dumped inside industries' premises. The Addis Ababa Abattoirs is a good example of this. The mountain of bones piled up inside the compound of the slaughterhouse, located on a major thoroughfare, is dreadfully stinky and aesthetically repulsive. Those industries that burn their solid wastes in the open inside their complex are no less obnoxious. The nauseating smoke released from such burning can be sensed several kilometers away. Some industrial wastes that are collected by municipal solid waste trucks are disposed of at the Ayer Tena municipal waste dumpsite. Hazardous industrial solid wastes thus end up in the city's dump ground that is supposed to receive only non-hazardous municipal solid wastes. People who make less than a subsistence living by scavenging salvageable materials from the dumpsite are daily exposed to dangerous industrial wastes.

The level of industrial waste generated in Ethiopia and the resulting pollution is difficult to gauge due to lack of data and adequate surveys conducted in the field. However, limited surveys in and around Addis Ababa do indicate that though the industrial sector is relatively young, it is already a culprit of environmental pollution. The listing below provides general information on the sources and types of solid waste and liquid wastes generated at industrial sites in Addis Ababa by industrial groups (EPA, 2001; CSA, 2000; CSA, 2001b; Tchobanoglous, et. al., 1993: 46-47). These wastes may not include all the hazardous or toxic wastes that each industry generates. Such hazardous or toxic wastes may include those that are highly flammable (for example, solvents used in chemical industries), highly reactive (capable of exploding or generating toxic gas when coming into contact with water or other chemicals), pathogenic (such as hospital wastes and sewage sludge), lethal (like cyanide, arsenic and many heavy metals), or carcinogenic (for instance, heavy metallic compounds).

All these wastes require special care in handling, storing, transporting, and disposing, but unfortunately very few industries in the country classify their toxic wastes and dispose of them properly. Industries in Addis Ababa use a wide variety of chemicals, perhaps as many as 800 or more (AACG, 2000: 16-17).

*Tanning industries* Waste-generating processes involve leather tanning, dressing, and finishing, manufacturing of handbags, luggage, and footwear. These industries are by far the worst polluters. Water is used extensively in tanning hides, either for washing or as a solvent for chemicals. Liquid waste from tanning skin and hides contains chrome, sulfides, ammonium, salts, and chlorides. The industry yields solid waste from dehairing, fleshing, and trimming of hides and skins. Solid waste products of animal origin are potent pollutants in water and are also highly odorous when they decompose in their solid form. Leather processing generates a heavily contaminated liquid waste, considered to be extremely hazardous due to the abundant presence of heavy metals.

*Textile industries* Waste-generating processes include spinning, weaving and finishing of textiles, manufacture of cordage, rope, twine and netting, and knitting mills. Expected specific wastes may include effluents heavy with organic compound. Textile liquid waste generation depends on the type of fibers (wool, cotton, synthetics, or a combination of these) and processes involved (purified and then bleached, spun, woven, or dyed). Purification and wet chemical processing of cotton generates mainly organic and chemical pollutants. Bleaching and dyeing often produce toxic chemicals. Liquid waste from scouring, mercerizing, bleaching, and dyeing process of textiles contain NaOH, peroxides, aluminum compounds, and dyestuffs.

*Beverage industries* Waste-generating processes involve distilling, rectifying and blending of spirits, manufacture of wines, malt liquors and malt, soft drinks and production of mineral waters. Most of the wastes from these industries are slightly acidic; they contain proteins and carbohydrates, yeast in suspension, and some detergents. Liquid waste is derived from barley steeping, pressed grains, hop and yeast recovery, cooling water discharge, spillage, and wash water.

*Food processing industries* Waste-generating processes involve manufacture, processing and preserving of meat, vegetables and fruits, manufacture of vegetables and animal oils, fats, dairy products, animal feeds, bakery products, sugar and sugar confectionary, and cereals. Expected specific wastes include insoluble organic matters: meats, fats, bones, offal, vegetables, fruits, cereals, and shells.

*Chemical industries* Waste-generating processes involve manufacture of paints, varnishes and lacquers, pharmaceuticals, medicinal chemicals and botanical products, soaps, cosmetics, perfumes and other cleaning preparations. Expected specific wastes may include organic and inorganic chemicals, strong acids and bases, metals, plastics, oils, paints, spent solvents, reactive wastes, and pigment.

**Table 5.2  Methods of Wastewater and Solid Waste Disposal for Selected Industries in Addis Ababa**

| Factory Name | Type of treatment for wastewater | Wastewater disposed into | Solid waste disposal |
|---|---|---|---|
| Awash Tannery | New facility under construction | Little Akaki River | Municipal pick-up |
| Ethiopickling and Tanning Factory | None | Little Akaki River | Municipal pick-up |
| Addis Ababa Tannery | None | Leku & Gafarssa river | Dumped inside factory |
| Teramaj Edible Oil Factory | None | Little Akaki River | Incinerated in factory dump site |
| Akaki Textile Factory | Partial treatment | Great Akaki river | Dumped into river Dumped inside compound |
| Awash Wineries, 1 & 2 | None | Little Akaki River Mekanisa River | Dumped inside compound |
| National Alcohol & Liquor Factory | None | Little Akaki River | Dumped inside compound |
| Edget Yarn and Sewing Thread Factory | Modern & full treatment facility | Tributary of Little Akaki | Pick-up Municipal |
| St. George Brewery | None | Tributary of Little Akaki | Dumped inside compound |
| Addis Ababa Abattoirs | None | Little Akaki River | Dumped inside compound |
| East Africa Bottling | None | Little Akaki River | Dumped inside compound |
| Ethiopia Marble Industry | None | Little Akaki River | Dumped inside compound |
| National Tobacco Enterprise | None | Little Akaki River | Dumped inside compound |
| Ethiopia Meat Concentrate Factory | None | Little Akaki River | Dumped inside compound |

| | | | |
|---|---|---|---|
| Addis Ababa Bottle and Glass Factory | None | Tributary of Little Akaki River | Dumped inside compound |
| Equatorial Paint Factory | Settling tank | Open drainage | Dumped inside compound |
| Nifas Silk Paint Factory | None | Open drainage | Municipal pick-up |
| Gulele Soap Factory | Neutralization tank | Open ditch | Dumped & incinerated inside |
| Repi Soap Factory | None | Open drainage | Dumped & incinerated inside |
| Addis Car Battery | None | Open drainage | Dumped inside compound |
| Chora Oxygen and Acetylene Factory | None | Open drainage | Dumped inside compound |
| Alkyd Resin Factory | Non-operational treatment facility | Open ditch | Dumped inside compound |
| Addis Tyre Factory | None | Open drainage | Dumped inside compound |
| Dil Edible Oil Factory | None | Open drainage | Dumped inside compound |
| Adei Abeba Yarn factory | None | Open ditch | Dumped inside compound |
| MOHA Soft Drink factories, 1 and 2 | None | Open drainage | Dumped inside compound |
| East Africa Soap and Detergent Factory | None | Open drainage | Dumped inside compound |
| Kadisco Chemical Industry | Septic tank | Open drainage | Dumped & incinerated inside |
| Ethiopian Pharmaceuticals Factory | None | Open drainage | Municipal pick-up |
| Summit Beverage Pvt.Ltd | Septic tank | Open ditch | Dumped inside compound |

Source: Addis Ababa City Administration Environmental Protection Bureau (2000), *Proceedings of the Workshop on the State of Industrial Pollution in Addis Ababa and Possible Remedies*, Addis Ababa: Addis Ababa City Administration, p. 2.

*Iron and steel (metal) industries* The wastes from iron and steel processing may be categorized as: *machining*, metal particles from machining usually mixed with lubricants; *degreasing*, metals mostly in solution with cyanides, alkalis, and solvents; *pickling*, acids with metals and metallic oxides in solution; *dipping*, alkalis with sodium carbonate, dichromate plus metals; *polishing*, particles of metals and abrasives together; *electrochemical or chemical brightening and smoothing*, acids, mainly sulfuric, phosphoric, chromic and nitric with metals in solution; *cleaning*, hot alkalis with detergents, cyanides and dilute acids plus metals in solution; *platting*, acids, cyanides, chromium, salts, pyrophosphates, sulfamates and fluroborates plus metals in solution; and *anodizing*, chromium, cobalt, nickel, and manganese in solution (James and McDougal, 1995, 139). Air pollution is characteristic of foundry operations. Electroplating is a primary pollutant of water, however. Polluting activities include cleaning and the electroplating process itself. Electro-plating uses toxic chemicals. Liquid waste from air scrubbers has high-suspended solids and a low pH. Acidic water is especially common from the pickling area.

*Pulp and paper manufacturing industries* Waste-generating processes include paper manufacture, conversion of paper and paperboard, manufacture of paperboard boxes and containers. Expected specific wastes include bleach residues and organic load, paper and fiber residues, chemicals, ignitable solvents, paper coatings and fillers, inks, glues, and fasteners.

*Printing industries* Waste-generating processes involve plate preparation, photo processing, printing, and cleanup. Specific wastes produced may include such wastes as inks, spent electroplating wastes, ink sludge containing heavy metals, spent solvents, and heavy metal solutions.

*Machinery and equipment industries* Waste-generating processes involve manufacture of pumps, compressors, taps and valves, ovens, furnaces and furnace burners, general-purpose machinery, equipment for construction and food and beverage processing, and primary cells and batteries. Expected specific wastes include scrap metals, cores, slag, plastics, rubbers, paints, and solvents.

*Wood and wood products* Waste-generating processes stem from sawmills, millwork plants, and miscellaneous wood products. Specific wastes include sawdust, wood preservatives, paints, varnishes, scrap wood, and shaving.

*Rubber and plastic industries* manufacture of fabricated rubber and plastic products. Specific wastes include scrap rubber and plastics, curing compounds, and dyes.

*Non-metallic mineral industries* Manufacture of structural clay products, cement, lime, plaster, and concrete products, and miscellaneous non-mineral products. Specific wastes produced include cement, clay, ceramics, stone paper, abrasives, dust and particulate air pollutants, and air pollution from combustion of fuels.

*Fabricated metal products industries* Waste-generating processes involve manufacture of metal cans, hand tools, general hardware, pluming fixtures, fabricated structural products, wire, and coating of metal. Specific wastes include metals, ceramics, sand, slag, coatings, solvents, lubricants, and pickling liquor.

*Vehicle maintenance shops* Waste may include ignitable wastes, metal scrapes, and paint wastes containing heavy metals, used lead acid batteries, and spent solvents.

## Effects of Industrial Liquid Waste on Surface Water

Industrial and household liquid wastes contain biodegradable organics composed principally of proteins, carbohydrates, and fats. Biodegradable organics in liquid waste are measured most commonly in terms of biochemical oxygen demand (BOD), chemical oxygen demand (COD), and dissolved oxygen (DO). If discharged untreated into the environment, their biological stabilization can lead to the depletion of natural oxygen resources and to the development of septic conditions. The pathogenic organisms in biodegradable organic-rich liquid waste can also transmit communicable diseases (Tchobanoglous and Burton, 1991: 49).

Low DO, high BOD, and high coliform counts all characterize most surface waters in and around the city of Addis Ababa. DO is perhaps the most important measure of water quality. DO is the amount of oxygen that is found dissolved in water (Koren, 1991). The levels of DO vary with temperature and altitude. 'Cold water holds more oxygen at a higher altitude than warm water, and water holds less oxygen at a higher altitude. Thermal discharges (such as water used to cool machinery in a manufacturing industry or a power plant) raise the temperature of water and lower its oxygen content.' Water has 14.60 mg/l of DO at zero degree centigrade, 11.27 mg/l of DO at 10 degrees centigrade, 9.07 mg/l at 20 degrees centigrade, 7.54 mg/l at 30 degrees centigrade, and 6.41 mg/l at 40 degrees centigrade (Spellman and Drinan, 2000: 197-98). Dissolved oxygen is essential for fish and other aquatic life to carry out all biological functions. Without dissolved oxygen, streams and lakes become devoid of aquatic organisms. Natural systems generally have waters with 8 to 12 parts per million of dissolved oxygen. This is considered 100 percent oxygen. Most fish species need at least 50 percent DO to survive. When sewage and industrial effluents are allowed into natural waterway, bacteria decomposition of their organic components use dissolved oxygen. As a result, dissolved oxygen levels can drop sharply threatening the natural ecology of the system. Fish die, algae blooms, and certain water weeds pullulate as a result of drops in DO caused by the introduction of sewage and organic industrial wastes into natural waterways (Koren, 1991). Discharging heated water into lakes, rivers, or streams can endanger aquatic life because water cannot retain dissolved oxygen and other gases in high temperatures that quicken biological and chemical processes (Moeller, 1997: 155).

Measuring the rate at which dissolved oxygen is used is as important as determining the amount of dissolved oxygen. The rate of oxygen use is referred to as BOD (Vesilind, et. al., 1990: 49).[1] It is a measure of the amount of oxygen required by bacteria to stabilize decomposable organic matter (sewage, in this case) under aerobic conditions (Teka, 1984).[2] Surface water that has enough dissolved oxygen to meet the biological oxygen demand of incoming sewage and maintain aerobic conditions essential for aquatic life is a healthy system. However, the BOD imposed by the organic decomposition of sewage often reduces the total dissolved oxygen in a system and threatens its native aquatic life. Under anaerobic conditions of decomposition algal blooms and putrefaction (characterized by offensive odors) can occur (Moore, 1999).

COD is used to determine how many compounds harmful to biological life are present in industrial and municipal waste (Tchobanoglous and Burton, 1991: 82). COD values are always higher than BOD values because almost all organics can be chemically oxidized in the COD test and only some can be biologically oxidized during the BOD test (Vesilind, et. al., 1990: 54).

Another significant water quality measure is the chloride concentration. Chloride concentration in Addis Ababa surface waters has been found to be high. Industrial and domestic liquid wastes discharged to surface waters are a major source of chlorides. 'Human excreta, for example, contain about 6 grams of chlorides per person per day' (Tchobanoglous and Burton, 1991:85). High chloride concentrations indicate that the surface water is being used for the disposal of domestic sewage and industrial effluents.

High levels of nitrogen and phosphorus compounds enter surface waters in domestic and industrial discharges and storm water run-offs. For instance, municipal liquid wastes may contain between 4 and 15 mg/l of phosphorus (Tchobanoglous and Burton, 1991: 86). The elements nitrogen and phosphorus are essential nutrients for growth of plants. When discharged into the aquatic environment these nutrients, or bio-stimulants, can stimulate the growth of undesirable aquatic life. As algae and aquatic organisms die, they drop to the bottom of the lake. Aerobic bacteria use all available dissolved oxygen to decompose this material. Consequently, all aerobic aquatic life disappears. When discharged in large quantities on land, they can lead to the pollution of ground water as well. Suspended solids can also result in the development of sludge deposits and anaerobic conditions when untreated liquid waste is discharged in the aquatic environment (Tchobanoglous and Burton, 1991: 49). Toxic inorganic compounds such as lead, chromium, and cyanides found in industrial wastes, particularly in metal-plating wastes, have been found in surface waters as well.

These are toxic in varying degrees to microorganisms. Dissolved inorganic contaminants in liquid waste are major health concerns. A few years ago, beef exported to Europe from a meatpacking firm in Addis Ababa was found to contain a high level of sulphide. The sulphide found in the meat was traced back to beef cattle that consumed water from a river polluted by wastes discharged from tanneries (EPA/UNIDO, 2001c: 35).

Pollutants in industrial liquid waste can be organic, inorganic, and radioactive substances. Organic compounds tend to cause changes in odor, taste, and color of water, whereas inorganic compounds often increase turbidity and foam and affect odor of water. Bacteria, viruses, and other plant and animal life can also grow in industrial liquid waste and affect water quality. In Addis Ababa, the food, textiles, tanneries, and chemical and metal processing industries are the most pollutants.[3] For instance, the 'pollution arising from beverage processing effluents alone has recorded a high level of BOD reaching 12,169 mg/l from a single factory discharge point, suspended solids of 429 mg/l, nitrate discharge of 13.5 mg/l, and sulfate of 278 mg/l from various locations' (EPA/UNIDO, 2001b: 5). Brewery liquid wastes, which contain beer, malt, and yeast, may have as much as 1,800 mg/l of BOD, 3,000 mg/l of COD, 1,750 mg/l of total solids, 800 mg/l of suspended solids, a pH of 7.5, an alkalinity of 160 mg/l as $CaCO_3$, and they have a high temperature (Polprasert, 1996: 40). The effluents from the Addis Ababa slaughterhouse are usually heavily polluted with solids, manure, and fat and protein-rich compounds. The wastes originate from slaughtering, hide stripping, stomach removal, trimming, processing, and clean-up operations. Slaughterhouse liquid waste is estimated to contain 360-1,880 mg/l of BOD, 489-1,425 mg/l of suspended solids, 36-510 mg/l of organic N, 88-440 mg/l of grease, 190-5,690 mg/l of chloride as Cl, an upper range of 2,000-4,900 mg/l of COD, and a pH of 6.5-8.4 (Polprasert, 1996: 45).

Textile industries have also recorded high levels of suspended solids reaching 160 mg/l, dissolved solids of 966 mg/l, nitrate of 50 mg/l, sulfate of 58 mg/l, and chloride of 116 mg/l from a single textile factory discharge point. Tanneries have the worst record of pollutant liquid waste discharge. High levels of pollutants were recorded from a single factory location: 1,350 mg/l of suspended solids, 17,425 mg/l of dissolved solids, 18,750 mg/l of chloride, 113 mg/l of nitrate, and 3,276 mg/l of sulfate and 2,782 kg load of BOD (EPA/UNIDO, 2001b: 5). The Addis Ababa Water and Supply Authority's rules require liquid waste discharge into public drainage systems contain no more than 2,000 mg/l of total dissolved solids, 2,000 mg/l of total suspended solids, and 650 mg/l of biochemical oxygen demand. The temperature of the liquid waste should not exceed 50 degrees Celsius, the pH should be between 5 and 10, and the liquid waste ought to contain no floating particles or substances that may cause the system to malfunction (EPA/UNIDO, 2001a: 50).

Haphazard disposal of industrial wastes has resulted in the contamination of farms around the industrial town of Akaki. Sample soils tested for heavy metals have been found to contain high levels of lead, nickel, copper, chromium, mercury, and cobalt. Chromium and nickel contents in the soil have reached toxic levels (Addis Ababa City Administration, 2000: 30). Hydrocarbon solvents that are widely used in industries as cleaners, degreasing agents, constituents of glues and paints, paint or coating removers, and other industrial contaminants have also found their way into groundwater sources and farms in the area.

**River Pollution**

Two major rivers—the Little and Great Akaki—and Lagadadi, Gafarssa, Dire, and Aba Samuel reservoirs form the main surface waters in and around the city of Addis Ababa. The two rivers originate from the Entoto Mountain range and are fed by several tributaries as they traverse through the city. Tributaries such as the Ginfile, Qabbana, Qachane, Qurtume, and Yeka flow into the Great Akaki River while Fereja, Buhe, Mekanisa, and other smaller steams drain into the Little Akaki. Both the Little and Great Akaki rivers finally end up discharging their waters into Lake Aba Samuel about 40 kilometers south of the city. All the surface waters in the city receive a wide variety of pollutants from a large number of point sources and from dispersed non-point sources. Industrial effluents compromise the water quality of surface waters in the city. Pinpointing the most severe pollution sources is difficult because the city does not require industries and other polluters to record their quantities of discharge and their chemical composition. Urban storm run-off washes away a variety of contaminants that end up in the city's rivers. Among the most significant sources of contaminants carried by urban storm water include eroded soil, decayed vegetation, solid waste, animal droppings, metals, construction debris, vehicular exhaust residue, and human feces.

Of all the city's surface waters, Little Akaki River is by far the most polluted river because several large industries line its banks and tributaries upstream in the western portion of Addis Ababa. The river water of the Akaki has many traditional uses including domestic consumption for laundering, bathing, cooking, drinking, sand mining, irrigation, and drinking water for cattle. Fifty years ago, the river's water was reasonably clean and safe for use; however, the water quality of the Little Akaki is quickly deteriorating. This deteriorating water quality has not stopped a significant portion of the urban and rural population from continuing their traditional uses of the river water. Those that rely on the river water include the urban poor, the homeless, rural farmers, cattle keepers, and sand miners. People recognize the decline in the water quality of the river: they notice the sickening stench of the water, its dark color, and the waste load of medical refuse and raw sewage. Still, their financial destitution forces them to use the river. Most lack access to clean drinking water, nor do they have the resources to filter the river water, drill for clean groundwater, or properly dispose of waste (Berhe, et. al., 2000: 17-19).

The river shows little evidence of pollution upstream. The BOD and COD values in upstream samples showed only 8-11 mg/l and with high DO in the water (Alemayehu, 2001: 98). However, once it enters the city a vivid and fast deterioration of its water becomes quite noticeable, rapidly reducing its quality to no better than that of raw sewage (EPA/UNIDO, 2001b: 6). Analysis of water samples taken from the middle course of the Little Akaki River contained 301 mg/l of BOD, 621 mg/l of COD, and zero dissolved oxygen. At the inlet to Aba Samuel Lake, water samples from the same river contained 321 mg/l of BOD, 708 mg/l of COD, and no dissolved oxygen (EPA, 2001; Alemayehu, 2001: 100). The primary sources of pollutants are industrial effluents and domestic sewage (point sources) and storm water run-off (non-point sources).

The upper catchment of the Akaki River includes three *weredas* of the city. Several industries that produce leather products, glass, oil, marble, and detergents are located along or in proximity to the river. It receives far more industrial effluents, sewage, and solid wastes than the Great Akaki River that flows through mostly residential and commercial districts in the northeastern and southeastern parts of the city. The latter river receives little industrial effluents until it traverses the industrial town of Akaki.

The middle catchment of Little Akaki comprises ten *weredas*. Several tributaries (such as Fereja, Buhe, and Mekanisa) feed into the middle course of the river, and each transports varying amounts of solid wastes, sewage, and industrial effluents. Fereja and Buhe, which flow through some of the densest residential and commercial districts of the city including Merkato, Ehil Berenda, Geja Safar, and Lideta, comprise the most polluted tributaries in the city.

Concentration of large- and medium-sized industrial establishments is found along the middle to lower course of the river, including industries that produce food and beverage (10 factories), leather goods (5), detergents (3), textiles (3), paints (2), chemicals (2), glasses, tires, pharmaceuticals, and slaughterhouse (one each). Because most of these industries have been built on the river's edge in order to use it as a waste receptacle, this section of the river receives the heaviest amounts of sewage, solid wastes, and untreated industrial effluents. Those industries that are not on the riverbank dispose of their wastes in ditches and municipal drainage systems, which reach the river indirectly.

Before it continues its journey to Aba Samuel Lake, the river receives its last pollutants from a point source at Kaliti, the Kaliti Sewage Treatment Plant. The plant discharges both treated and untreated sewage into the river. Some of the untreated sewage that finds its way into the river is the sewage that arrives every day by vacuum truck from the city. Because the treatment plant cannot process all of the raw sewage that is brought by vacuum trucks, some of it is released into the river. Past the town of Kaliti, the Little Akaki River meanders through farming communities and becomes the source of water for rural households and cattle. Communities living along the river seem to understand the dangers but continue to use the water because they cannot afford water from other sources.

In 1998/1999, the water quality of Little Akaki River was assessed by testing the water's BOD, DO, and COD, and by measuring concentrations of nitrogen, phosphate, suspended solids, *E. coli*, and coliform. These properties were measured at 22 sites along the river beginning at its source and continuing down through the city of Addis Ababa. Measurements were taken once during each season including the dry season, the wet season, and the *Belg* (little rain) season.

Overall the results suggest that the river is severely polluted and has surpassed a level that natural processes can manage. Unacceptable levels of BOD, DO, Nitrate, Phosphate, *E. coli*, and coliform in all three seasons indicate that the river is severely polluted year round, reaching the highest level of pollution in the middle catchments directly downstream from most of Addis Ababa's industries. Depleted oxygen levels during the dry season prevent most life forms from inhabiting the river at this section, while excess nutrients such as nitrogen and phosphorous cause similar oxygen depletions through eutrophication at the Aba Samuel dam, killing fish populations.

Based upon these results and the concentrations of *E. coli* and coliform in particular, the river water is highly unsafe for domestic use. No one knows whether the high concentrations of *E. coli*, coliform, nutrients, and BOD of the irrigated river water will pose a health risk to those who eat produce grown with this water source. We must conduct more research in order to make accurate risk assessments. Measuring heavy metal concentrations in soils irrigated with river water and in cattle that drink this water would be an important step in assessing the problem in conjunction with analysis of the sewage effluent direct from their source industries (Berhe, et. al., 2000: 66-67).

## Air Pollution

Air pollution is the sustained presence of chemicals in the atmosphere in sufficient quantity to harm living organisms and inorganic matter alike (Miller, 1998: 464-65). Air pollutants may exist in gaseous or particulate form. Gaseous pollutants include substances such as carbon monoxide, hydrocarbons, hydrogen, sulfide, nitrogen oxides, ozone, and other oxidants, and sulfur oxides (Table 5.3).

Particulate air pollutants are diverse in chemical composition and size, and they include both solid and liquid droplets. Particulate pollutants can be classified as *dust* (solid particles from coal, ash, cement, sawdust, sandblasting; *fume* (a solid particle resulting from condensation of vapors via sublimation, distillation, calcinations, or chemical processes); *smoke* (solid particles resulting from incomplete combustion of carbonaceous minerals; *mist* (a liquid particle formed by the condensation of vapor and by chemical reaction as well; and *spray* (a liquid particle formed by the atomization of a parent liquid) (Vesilind, et. al., 1990: 245-47).

Natural processes emit many pollutants. For instance, dust from volcanic eruptions, smoke particles from forest fires, fungus spores, and salt spray occur naturally. Examples of naturally forming gaseous pollutants may include sulfur dioxide from geysers, carbon monoxide resulting from plant and animal respiration, and 'sulfur-containing amino acids by bacterial action, nitrogen oxides, and methane' (Vesilind, et. al., 1990: 249). However, human activities are directly responsible for the vast volume of pollutants in and around agglomerated human settlements.

Most of the air pollution in Addis Ababa originates from three major sources: motor vehicle exhaust, industries, and energy use.[4] Air pollution from motor vehicles in the country is not anywhere near the magnitude reached in the developed world. However, the contribution to air pollution from motor vehicle emissions is growing very rapidly. Motor vehicle emissions cause one and a half to three times more pollution than the emission of industrial activities (EPA/UNIDO, 2001b: 7). Vehicle exhaust contains nitrogen oxides, carbon dioxide, volatile organic compounds, and particulate matter. Although no data on these pollutants exist, it is likely that the increasing number of motor vehicles over the past two decades or so has increased the volume of these pollutants. In 1980, there were fewer than 70,000 motor vehicles in the nation as a whole. In 1996, there were 140,000 registered motor vehicles in the streets of Addis Ababa alone (Malifu, 2001: 144) and this number grows significantly every year.

Motor vehicle emissions depend on several factors. Emissions depend on fuel used to power vehicles. Petrol-powered vehicles emit more CO and volatile organic compounds (VOCs) and are less fuel-efficient than similar diesel-powered vehicles. Newer vehicles with pollution control emit fewer pollutants than older vehicles with older engines. Poorly maintained vehicles consume more fuel and emit higher levels of CO and VOCs than those regularly serviced. Most vehicles in use in Addis Ababa are old and poorly maintained and thus the resulting pollution is huge. Traffic congestion increases emission, for stop-start driving conditions generate more emissions than normal traffic flow. Emissions of CO and VOCs from vehicle exhaust are particularly high when cars are first driven from a cold start. Emissions are higher for petrol cars than diesel cars (Holman, 1995: 295-99). The leaded gasoline sold in the city increases the atmospheric concentration of lead in the city.

Many industrial operations in Addis Ababa emit significant quantities of air pollutants. Practically every industrial process is a potential source of smoke, dust, or aerosol emissions. The smoke released into the air from industries such as the Addis Ababa Tire factory contains nitrogen oxide, sulfur dioxide, and carbon monoxide (Aberra, 2001: 183).[5] Industries that use solvent, in paints and adhesives are significant sources of hydrocarbons (volatile organic compounds). When sulfur dioxide ($SO_2$) from industries and nitric oxide from vehicle exhaust mix in the atmosphere, they transform to sulfuric acid ($H_2SO_2$) and nitric acid ($HNO_3$). These chemicals descend to the land surface as dry acid deposition (sulfur dioxide and particles of sulfate and nitrate salts) or wet acid deposition

**Table 5.3  Gaseous Air Pollutants**

| Name | Formula | Properties of Importance | Significance as Air Pollutant |
|---|---|---|---|
| Sulfur dioxide | $SO_2$ | Colorless gas, intense choking odor, highly soluble in water to form sulfurous acid $H_2SO_3$ | Damage to vegetation, property, and health |
| Sulfur trioxide | $SO_3$ | Soluble in water to form sulfuric acid $H_2SO_4$ | Highly corrosive |
| Hydrogen sulfides | $H_2S$ | Rotten egg odor at low concentrations, odorless at high concentrations | Highly poisonous |
| Nitrous oxide | $N_2O$ | Colorless gas, used as carrier gas in aerosol bottles | Relatively inert, not produced in combustion |
| Nitric oxide | $NO$ | Colorless gas | Produced during high temperature, high-pressure combustion. Oxidizes to $NO_2$ |
| Nitrogen dioxide | $NO_2$ | Brown to orange gas | Major component in the formation of photochemical smog |
| Carbon monoxide | $CO$ | Colorless and odorless | Product of incomplete combustion. Poisonous |
| Carbon dioxide | $CO_2$ | Colorless and odorless | Formed during complete combustion. Possible effects in producing changes in global climate |
| Ozone | $O_3$ | Highly reactive | Damage to vegetation and property. Produced mainly during the formation of photochemical smog |

| Hydrocarbons | HC | Many | Some hydrocarbons are emitted from automobiles and industries, others are formed in the atmosphere |
| Hydrogen fluoride | HF | Colorless, acrid gas | Product of aluminum refining, causes fluorosis in cattle |

Source: T. Aarne Vesilind, James Peirce, and Rugh Weiner (1990), *Environmental Pollution and Control*, London: Butterworth-Heinemann, p. 246.

(droplets of $H_2SO_4$ and $HNO_3$) dissolved in rain. This form of pollution is known as acid deposition and has many detrimental ecological effects. Acid deposition is harmful to both the terrestrial and aquatic systems, especially when the pH falls below 5.5. It also poses a risk to human health, contributing to respiratory diseases such as asthma and bronchitis (Miller, 1998: 470-71).

Globally, energy conversion (fossil fuels and biomass burning, for example) is the largest source of air pollution. Energy related activities cause '90 percent of the total sulphur dioxide ($SO_2$) and lead (Pb), 85 percent of nitrogen oxides (NO), 55-80 percent of the carbon dioxide ($CO_2$), and 30-50 percent of the carbon monoxide (CO) emissions into the atmosphere' (EPA/UNIDO, 2001b: 44). Each phase of the energy life sequence—exploration, extraction, production, transportation, conversion, and final use—produces pollutants that are perilous to human health and the environment. In Ethiopia, the burning of biomass fuels is a major contributor to air pollution in urban areas. Most of the pollutants come from households burning low-quality fuels that are important domestic energy sources. Urban households use one of several of the following fuels: firewood, crop residues, leaves, charcoal, kerosene, LPG, and electricity. The higher the energy source, the less polluting the fuel. Ethiopia's emission of carbon dioxide is estimated at 5.5 million metric tons per year (World Bank, 2003: 148).

The Addis Ababa waste dumpsite also produces methane ($CH_4$), a greenhouse gas, considered to be much more potent than carbon dioxide in causing global warming.[6] No reliable data is available for the amount of methane generated by the dump, but it is well known that the decomposition of dead organic matter by anaerobic bacteria yields methane (Miller, 1998: 366). Anaerobic decomposition of organic waste at the dumpsite may also produce toxic hydrogen sulfide gas and smog, forming volatile organic compounds that escape into the air. Another source of methane gas is cattle waste from over 4,000 small-scale unregulated dairy enterprises located in and around Addis Ababa (Gebeyehu, 1996: 11-13). 'Nitrogen in manure escapes into the atmosphere as gaseous ammonia ($NH_3$), a pollutant that contributes to acid deposition' (Miller,

1998: 600). A study by the National Meteorological Service Agency indicates that the amount of methane generated from the municipal waste dump in Addis Ababa in 1998 was 9.39 Gg (Giga gram) (Malifu, 2001: 144).

Information on the health impacts of urban air pollutants in Ethiopia is non-existent. However, several studies from other parts of the world show that most air pollutants have detrimental effects on health. Exposure to high levels of *sulfur dioxide* can cause constriction of the airways in healthy individuals and severe restriction in asthmatic individuals. Long-term exposure causes a condition of chronic bronchitis. *Nitrogen oxides* 'can irritate the lungs, aggravate the condition of people suffering from asthma or chronic bronchitis, cause conditions similar to chronic bronchitis and emphysema, and increase susceptibility to respiratory infection, such as the flu and common cold' (Miller, 1998: 480). *Carbon monoxide*, which has no color, smell, or taste, wields its toxic effect by binding to hemoglobin and thereby reducing the oxygen capacity of the blood.

High doses of CO can result in death 'due to cerebral and cardiac hypoxia' and exposure to low level concentrations 'may affect higher cerebral function, heart function and exercise capacity, all of which are sensitive to lowered blood oxygen content' (Walters and Ayers, 1995: 258). *Suspended particulate matters* are harmful because they are small enough to penetrate the respiratory system. Inhaling them 'aggravates bronchitis and asthma and long-term exposure can contribute to development of chronic respiratory disease and cancer' (Miller, 1998: 479). A comprehensive study done in 36 Indian cities in 1995 indicates that suspended particulate matters (SPM) were responsible for high levels of death and morbidity. Globally, outdoor air pollution causes between 200,000 and 750,000 deaths (EPA/UNIDO, 2001b: 7).

**Indoor Air Pollution**

Domestic functions use large amounts of energy. Some of this energy comes from fuels which, for most people in the developing nations, consist of biomass sources. Many health problems in Ethiopia result from indoor pollution from the use of low-quality fuels for cooking and heating. This problem is by no means limited to rural households. Traditional biomass fuels—wood, charcoal, sawdust, animal dung, leaves, or crop residues—are major sources of energy for cooking in both rural and urban areas in Ethiopia. In 1998, about 98 percent of rural households and 80 percent of urban households relied on biomass fuels for domestic energy (CSA, 1999: 162). In 1995, residents of Addis Ababa consumed 692,375 tons of fuelwood, 121,400 tons of charcoal, 24,730 tons of crop residues, and 22,435 tons of dung. Overall, biomass fuels accounted for nearly one-third of

**Table 5.4** **Types of Cooking Fuel Used by Households in Selected Urban Centers, 1998**

| Selected Urban Centers | Types of Cooking Fuel | | | | | | |
|---|---|---|---|---|---|---|---|
| | Collected Firewood | Purchased Firewood | Charcoal | Butane Gas | Electricity | Crop Residue / Dung / Saw dust | Others |
| | (%) | (%) | (%) | (%) | (%) | (%) | (%) |
| Mekele | 2.4 | 58.6 | 5.2 | 9.6 | 5.4 | 3.1 | 6 |
| Assayita | 21.6 | 54.4 | 18.5 | 0.9 | - | 1.7 | 2.8 |
| Gondar | 9.5 | 48.6 | 19.1 | 1.5 | - | 15.9 | 4.4 |
| Bahir Dar | 10.5 | 56.5 | 10.6 | 4.5 | 1.2 | 3.2 | 10.9 |
| Dessie | 9.6 | 64.6 | 1.5 | 14.2 | 0.4 | 1.9 | 2.7 |
| Jimma | 21.3 | 51.5 | 17.1 | 5.7 | 0.8 | - | 2.3 |
| Adama | 9.1 | 17.8 | 14.2 | 38.5 | 0.8 | 12.8 | 4.1 |
| Bishoftu | 7.1 | 24.5 | 2.3 | 34.6 | 1.4 | 23.5 | 2.0 |
| Jijiga | 15.2 | 28.8 | 45.9 | 4.9 | - | 0.8 | 2.5 |
| Asosa | 21.4 | 46.9 | 26.5 | 1.9 | 0.4 | - | 2.5 |
| Awasa | 7.9 | 50.7 | 4.7 | 26.2 | 3.8 | 0.8 | 2.1 |
| Gambela | 31.8 | 42.1 | 16 | 3 | - | - | 3.8 |
| Harari | 9 | 37.9 | 6.2 | 33.3 | 1 | 4.8 | 2.8 |
| Addis Ababa | 2.9 | 18 | 1.6 | 47.6 | 8.2 | 7.5 | 2.1 |
| Dire Dawa | 6.4 | 35 | 7.6 | 44.4 | 0.8 | - | 4.1 |
| Amhara other Urban | 21.3 | 64.2 | 1.5 | 2.9 | 1.1 | 5.4 | 2.9 |
| Oromiya other Urban | 18.8 | 58.9 | 6 | 6.4 | 0.5 | 6 | 2.0 |
| SNNPR other Urban | 14.9 | 73.4 | 3.5 | 3.6 | 0.6 | 0.7 | 3.4 |
| Other Urban | 19.6 | 60.4 | 6.4 | 3.4 | - | 2.3 | 7.9 |

Source: Central Statistical Authority (1999), *Report of the 1998 Welfare Monitoring Survey*, Statistical Bulletin No. 224, Addis Ababa: CSA, p. 178.

all cooking needs of the city in 1998 (Table 5.4). Commercial fuels, including propane gas, electricity and kerosene, met the remaining 57 percent. Per capita biomass fuel use in Addis Ababa shows a declining trend as kerosene for cooking and electricity for baking *injera* increasingly replace fuelwood. Kerosene and gas emit 10 to 100 times less respirable particulate than wood fuels (WHO, 1999b: 72). Biomass fuels are also used for lighting and space heating.

Indoor pollutants are not only a discomforting nuisance; they produce serious health risks. Incomplete combustion smoke from biomass fuels 'contains pollutants such as hydrocarbons, carbon monoxide, oxides of nitrogen, sulfur dioxide, aldehydes, dioxin, and organic compounds including polycyclic aromatic hydrocarbons that are considered to be toxic and carcinogenic' (WHO, 1999b: 15). Biomass fueled open fires produce a relatively high concentration of airborne pollutants. The indoors of most poor households in Addis Ababa and other urban areas are characterized by overcrowding and insufficient ventilation. When biomass fuel is burned in such an unventilated house, it produces smoke that slowly builds up and blankets the entire house, exposing household members to a wide range of harmful health conditions.

The direct human effects of indoor pollution depend on the duration of exposure and the age and health conditions of the people exposed. Women are exposed to high levels of smoke for longer periods of time as they are primarily responsible for household chores and food preparation, activities that keep them inside the house most of the time. A woman cooking with biomass fuels in a room that lacks ventilation 'is exposed to the equivalent of more than a hundred cigarettes a day' (Crewe, 1995: 93). The very young and the very old are also at greatest risk from frequent exposure to smoke inhalation. Very young children are at increased risk because of the lack of a well-developed defense mechanism. The elderly are more susceptible to air pollution than other adults because of their disease- weakened immune systems. Worldwide, about three-quarters of a billion children and women are exposed to severe air pollution from cooking fires (World Bank, 1998: 1). Burns resulting from the use of biomass fuels in open fire or stoves are also one of the leading causes of injury for women and small children. The production and collection of biomass fuel itself has adverse health impacts, especially for women. Gathering of fuel can result in trauma, bites from venomous snakes, spiders, leeches, and insects, allergic reactions, fungus infections, and severe fatigues. Processing and preparing animal dung cakes can result in fecal/oral/enteric infection and skin infection. Charcoal production can result in carbon monoxide/smoke poisoning, burns/trauma, and cataracts (Yassi, et. al., 2001: 316).

Few studies have been conducted in the country to determine the health effects of smoke-related indoor pollutants. But, evidence from elsewhere in developing countries shows that pollution generated by burning biomass fuels in a poorly ventilated house is extremely harmful to health. Ailments such as chronic obstructive lung disorders, acute respiratory infections, heart diseases, lung cancer, eye infections, skin burns and scalds, and impaired fetal development can occur. Developing countries are particularly affected. A number of studies in Africa and Asia suggest that respiratory infections in children are strongly associated with indoor use of solid fuels for cooking (WHO, 1999b: 72).[7] Acute respiratory infections are the most common cause of death in the developing countries, taking the lives of four to five million children under the age of five annually (Crewe, 1995: 92). A study in South Africa found that 'Zulu children living in homes with woodstoves were almost five times more likely to develop a respiratory infection severe enough to require hospitalization' (Kossove, 1982: 622-24).[8] Similarly, a study carried out in Gambia found that 'children carried on their mothers' back as they cooked over a smoky cook stove contracted pneumococcal infections—one of the most serious kinds of respiratory infections—at a rate 2.5 times higher than non-exposed children' (O'Dempsey, 1996).[9] Pregnant women in India who are exposed to high levels of indoor smoke have also been found to experience a 50 percent increase in stillbirths and low birth weight (Mavalankar and Gay, 1991). Another study in India and Nepal showed that cardiovascular disease is more prevalent among women who have been exposed to indoor pollutants (WHO, 1992b). Generally, acute respiratory infection is a disease of poverty associated with poor housing conditions, lack of access to healthcare, overcrowding, and poor hygiene.

It is estimated that poor health as a consequence of dependence on biomass fuel supplies affects at least half of the world's population. In most cases, though, this problem does not receive enough attention. Environmental problems tend to focus on other issues, such as global atmospheric pollution, industrial hazards, and nuclear waste. When assessing indoor pollution, pollution experts focus on pollution specific to the developed world, such as tobacco smoke and gas stoves. Gas emits only one-fifth of the pollution generated by wood stoves. The lack of attention to the problems of household energy management and indoor air pollution in poorly ventilated, unsafe kitchens stems from the fact that women constitute the chief victims (Crewe, 1995: 92-94).

**Agricultural Pollution**

Agriculture is a major non-point source of chemical fertilizers, pesticides, animal wastes, and other organic matter that affect water quality. Leachate of these chemical substances enters waterways via surface run-off, erosion, irrigation return flows, and precipitation. Fertilizers are particularly harmful because of their common use in state farms and peasant farms. Pesticides are also commonly used on large farms. Water pollution increases with increased use of these products.

Inorganic fertilizers are fast replacing the traditional use of manure and composts. Traditional pest control is losing to imported environmentally hazardous pesticides. There is a marked increase in the commercial use of inorganic fertilizers in Ethiopia thanks largely to the funding provided by the

Japanese Sasakawa Foundation. Inorganic fertilizers increased from 1,048,060 quintals in 1984/1985 (UNEP, 1992: 63-64) to 2,358,882 quintals in 1997/1998 to 4,329,660 quintals in 1999/2000 (CSA, 2001a: 123-25). Pesticide use increased from 146,775 kilograms and 150,310 liters in 1986/1987 to 868,489 kilograms and 477,799 liters in 1990/1991 (UNEP, 1992) to 3,122,200 kilograms/liters in 1997/1998 (EPA, 2001: 91). Areas under perennial irrigation schemes are especially prone to contamination as pollutants are added to the groundwater system year-round. This should particularly concern people living in areas where fertilizers and pesticides are most often used and where people rely upon groundwater for drinking water.

Large-scale irrigation schemes create excellent breeding grounds for mosquitoes and improve conditions that favor the propagation of snails and other disease vectors. Since the development of irrigation schemes in the upper Awash River Valley in the early 1960, there have been increases in the incidence of malaria and *Schistosomiasis*. For instance, the irrigation schemes have increased *Schistosoma mansoni* by improving the habitat for its *Bulinus* intermediate host. *Schistosomiasis* is a debilitating parasitic worm disease spread by water snails, causing chronic urinary tract disease, cirrhosis of the liver, and bladder cancer (WHO, 1999c: 10). *S. mansoni* affects pastoralists and migrant farm workers who use the canal water for bathing and washing clothes (Lo, et. al., 1988, 200).

Pesticides are a particularly harmful group of synthetic chemicals. They are composed of a wide range of different types of organic molecules that are used to kill pests, including insects, weeds, and fungal pathogens. The most serious environmental problems associated with pesticides are the contamination of surface and ground waters and the destruction of aquatic life. For decades, Ethiopia has imported tons of pesticides to boost food production for its hungry population. But recent findings show that some of these pesticides were never used. In 1998, a FAO funded inventory located over 1,500 tons of obsolete pesticides stored carelessly or inadequately at 420 locations around the country. The inventory found 386,102 kilograms in 181 locations in Oromiya; 186,171 kilograms in 102 locations in the Amhara State; 169,341 kilograms in 10 locations in Addis Ababa; 156,366 kilograms in 2 locations in Dire Dawa; 136,492 kilograms in 36 locations in SNNPR; 107,489 kilograms in 6 locations in the Afar State; 74,473 kilograms in 52 locations in Tigray; 59,595 kilograms in 9 locations in Benishangul-Gumuz; 12,589 kilograms in 2 locations in Gambela; and 8,301 kilograms in 2 locations in Somali (Megerssa, 2001: 7). These stored obsolete pesticides, considered to be among the most toxic chemicals, have contaminated the air, soil, and water for over 40 years and threaten the residents' health. The storage of the pesticides was found in appalling conditions: drums leak, sacks explode, and the stores in a horrible condition (Megerssa, 2001: 7). In some cases, stored obsolete pesticides have corroded, spilled out, and been washed down with rains to pollute nearby water sources (including leaking into the water table) that are used by domestic animals and households. A project funded by the USAID, Japan, Italy, Sweden, and the Netherlands was underway in 2000 to find ways to destroy these hazardous materials under the supervision of an international NGO, Pesticides Action Network (PAN-UK) (EPA, 2001: 2; Megerssa, 2001: 2).

Even as it tries to remove old pesticides, the country imports new ones. Pesticide use is still quite low in comparison to other developing countries; however, increasing numbers of farmers are using fertilizers. As import increases more and more people will be involved in handling it and hence a greater number of people will risk exposure. For example, exposure to pesticides causes damage to the brain, skin, and liver, and it produces chemical burns of the eye (Yassi, et. al., 2001: 276).[10] People in poor countries like Ethiopia are particularly prone to exposure through the eyes and skin, as pesticides are handled with bare hands. Pesticides may also be ingested through the consumption of contaminated food and water. Laborers who are frequently exposed to pesticides or their residue may face greater health risks than others. Pesticides are known to accumulate in the food chain—a phenomenon known as bioaccumulation. Low levels of concentration in lower organisms may have detrimental effects on higher organisms that feed on lower organisms (Chapman, 1998: 10).

## Conclusion

Increased industrial production, expanded extractive activities, and growing numbers of polluting motor vehicles will no doubt contribute to growing pressures on the urban environment. In the past, the general predisposition of the government has been to relax existing regulatory standards in an effort to entice foreign investors and to help domestic private enterprises to reduce their cost of production. Foreign or domestic investors are not required to submit environmental impact statements of their project when they apply for an investment permit, but the law demands that the investment activity 'comply with conditions stipulated in environmental protection laws' (EPA/UNIDO, 2001a: 22-23). Public Health Proclamation No. 200/2000 also prohibits industries from discharging untreated liquid waste into nearby rivers. However, the proclamation provides no monitoring or enforcement mechanisms. Neither does it provide minimum effluent thresholds or standards for determining environmental violations (EPA/UNIDO, 2001a: 37), nor the level of penalty that should be imposed on violating industries. A public Environmental Protection Authority does exist, but it has no power to regulate or enforce exiting environmental laws or impose penalties on offenders for breaking the laws.[11] Moreover, it is difficult to enforce existing environmental laws when the state owns many of the industries that break the laws.

Industrial waste disposal management requires an identification of the types of industrial wastes, full-scale implementation of treatment methods, and, ultimately, safe disposal of these wastes. However, few industries in Addis Ababa and other urban areas care to document the types and quantities of wastes they generate let alone treat and dispose of them safely. One might then say that it is time that EPA's proposed pollution control regulations are enacted into law and strictly enforced: no industries ought to exist without proper treatment plants and safe waste disposals, and those that pollute the environment ought to defray the cost of clean up. Unfortunately, things are often, but not invariably, more complicated in reality than in theory. It is difficult to imagine that a government that owns and runs a significant proportion of the major industries in the country would force itself to submit to its own environmental guidelines and standards or penalize itself upon failing to do so. A government cannot enforce the very laws that it violates. For instance, the law says that the Ministry of Trade and Industry has the legal responsibility to suspend licenses of commercial and industrial activities if and when failure to maintain health and environmental standards has been ascertained or confirmed by responsible government institutions (EPA/UNIDO: 2001a: 51). In spite of gross violations of environmental guidelines and regulations by practically every industry in the country, not a single industry has ever lost its license on account of such infringements.

## Notes

[1] The BOD test is usually run in the dark at 20 degrees centigrade for five days. This is defined as a five-day BOD, or the oxygen used in the first five days. The long period of time required to obtain results is a major limitation because 'the five-day period may or may not correspond to the point where the soluble organic matter that is present has been used' (George Tchobanoglous and Franklin L. Burton (1991), *Liquid Waste Engineering: Treatment, Disposal and Refuse*, New York: McGraw-Hill, p. 82).

[2] Aerobic bacteria is 'a type of bacteria that require free oxygen to carry out metabolic function' (Frank R. Spellman and Joanne Drinan (2000), *The Drinking Water Handbook*, Lancaster, PA: Technomic Publishing Co., p. 13).

[3] In 2000, the estimated industry share of emissions of organic water pollutants was 58.5 percent for food and beverages, 24.9 percent for textiles, 8.8 percent for paper and pulp, and 2.1 percent for wood (World Bank (2003), *World Development Indicators*, New York: Oxford University Press, p.140).

[4] Outside of Addis Ababa air pollution is not much of a problem simply because other urban areas have much smaller population and significantly lower ratios of motor vehicles to population.

[5] Edlam Aberra's study of the environmental impacts of the Addis Tire Factory (ATF) on an impoverished residential community in Addis Ababa that is located in the plant's immediate vicinity indicate that environmental hazards from industries within the city of Addis Ababa are of great concern to many residents. When residents were asked to describe the impact of pollution in their daily lives, nearly all mentioned the bad smell in the area. All those interviewed complained of the black soot emitted by the ATF coating the walls of their homes, clothes, and polluting their drinking water and food. Nearly all

those interviewed also complained about frequent nausea and breathing problems stemming from the stench of the emission while one-half said they suffer frequent headaches. All residents said the thick fog of pollution that settles on their community has forced them to dry their laundry only on Sundays when the plant is closed. Over three-quarters of the residents expressed their desire to move somewhere else but explained that housing shortages in the city and higher rental rates in other areas prevented them from making a move.

[6] Methane 'absorbs 20 to 30 times as much infrared radiation and is accumulating in the atmosphere nearly twice as fast as carbon dioxide' (William P. Cunningham and Barbara W. Saigo (2001), *Environmental Science: A Global Concern*, New York: McGraw Hill, p. 387).

[7] Studies have shown that domestic use of polluting 'fuels may facilitate the spread of acute respiratory infections by irritating the respiratory passage,...contributes to chronic obstructive lung diseases—a significant health problem among adult women'—and poses a risk factor for cancer as a result of a long-term exposure (B. H. Chen, C. J. Hong, and K. R. Smith (1990), 'Indoor air pollution in developing countries,' *World Health Statistics Quarterly*, 43: 127-138; Cited in McGranahan, Simon Levin, et. al. (1999), *Environmental Change and Human Health in Countries of Africa, the Caribbean and the Pacific*, Stockholm: The Stockholm Environmental Institute, p. 145).

[8] Cited in The World Resources Institute (1998), *World Resources: 1998-1999*, New York: The World Resources Institute, p. 66.

[9] Ibid.

[10] It is, however, important to note that the severity of health impacts 'from exposure to pesticides depends on the dose, the route of exposure, the type of pesticide, the absorption of the pesticide, and the health of the affected individual' (Annalee Yassi, et. al., 2001: 277).

[11] Since the early 1990s the government has put in place environmental policies to deal with environmental degradation and resource conservation in the country. While these policy efforts are commendable, very little has been achieved in dealing with the country's environmental problems owing largely to failure to translate policies into practice. The government has also signed four international treaties and agreements on environmental issues: *The Framework Convention on Climate Change* that 'aims to stabilize atmospheric concentrations of greenhouse gases at levels that will prevent human activities from interfering dangerously with global climate.' *The Vienna Convention for the Protection of the Ozone Layer* that 'aims to protect human health and the environment by promoting research on the effects of changes in the ozone layer and on alternative substances (such as substitutes for chlorofluorocarbons) and technologies, monitoring the ozone layer, and taking measures to control the activities that produce adverse effects.' *The Montreal Protocol for CFC Control* that 'requires countries to help protect the earth from excessive ultraviolet radiation by cutting chlorofluorocarbons consumptions.' *The Convention on Biological Diversity* that 'promotes conservation of biodiversity among nations through scientific and technological cooperation, access to financial and genetic resources, and

transfer of ecologically sound technologies' (World Bank (2003), *World Development Indicators*, New York: Oxford University Press, p. 193).

# Chapter 6

# The State of Health and Health Services

The Constitution of the World Health Organization (1946) and the Universal Declaration of Human Rights (1948) recognize health as a fundamental human right. The Constitution of the World Health Organization provides the most commonly quoted definition of health: 'a state of complete physical, mental and social well being and not merely the absence of a disease or infirmity' (WHO, 1948). Within the framework of this inclusive definition the World Health Organization promoted its 'The Health for All' declaration in the global health conference held in Alma Ata in 1978. The conference's resolution called on all governments, international organizations, and the global community to strive for 'the attainment by all people of the world by the year 2000 of a level of health that will permit them to lead a socially and economically productive life.' The conference also stressed that such a goal necessitates the allocation of adequate resources for health and health related needs. 'Health is only possible,' the conference declared, 'where resources are available to meet human needs and where the living and working environment is protected from life-threatening and health-threatening pollutants, pathogens, and physical hazards' (WHO, 2001a: 5).[1]

The year 2000 has come and gone, but the health status of the majority of the developing world remains poor. As Yassi and others (2001: 5) put it: 'Every year hundreds of millions of people suffer from respiratory and other diseases associated with indoor and outdoor air pollution. Hundreds of millions of people are exposed to unnecessary physical and chemical hazards in the workplace and living environment.... Four million infants and children die every year from diarrheal diseases, largely as a result of contaminated food or water. Hundreds of millions of people suffer from debilitating intestinal parasites. Three million people die from malaria every year while over 300 million are ill with it at any given time.[2] Three million people die each year from tuberculosis and 20 million are actively ill with it. Hundreds of millions suffer from poor nutrition.

The healthcare system in Ethiopia is appallingly inadequate. Only Addis Ababa and a few other major urban areas are relatively well serviced and the rural population has little or no access to basic healthcare facilities. This chapter provides a review of the state of the present health sector infrastructure and pattern of health services, health service resources, problems facing the health sector in Ethiopia, and the general public health policy environment.

**Modern Health Services: Historical Background**

The introduction and practice of modern medicine in Ethiopia began in 1896 with the establishment of the first hospital in Addis Ababa by the Russians. In 1899, Emperor Menelik II issued the first medical proclamation making vaccination against smallpox compulsory in Addis Ababa. The 1900s saw the establishments of a special Department of Public Hygiene and the first public hospital and pharmacy in the capital to serve members of the royal family and the nobility. Ethiopians went abroad for medical training; assistant medical personnel received training in mission and government hospitals in Addis Ababa (Buschkens and Slikkerveer, 1982: 46). On the eve of the Italian invasion in 1936, Ethiopia had about 12 hospitals and 30 clinics: some were run by the government; others, by foreign religious missions. The Italians brought to Ethiopia many doctors during their occupation (1935-1941) who served their soldiers and the Italian settler population. They introduced preventive medical programs to protect the health of their nationals. The Italians established a number of provincial hospitals and clinics and dispensaries in rural areas, but these facilities served very few people due to the ongoing war and popular resentment against the Italian occupation (Buschkens and Slikkerveer, 1982: 46-47). The pace of modern medicine increased after independence in 1941 and, toward the end of the 1940s, the country had a total of 106 doctors, 88 nurses, 46 hospitals, 3,400 hospital beds, 6 health centers, and 92 health stations (Desta and Meche, 1999: 13-18).

The establishment of the Ministry of Health (MOH) in 1948 ushered in the beginning of the expansion of health services elsewhere in the nation. In 1950 foreign physicians (mostly missionaries) served in hospitals in Addis Ababa and other major urban areas. A number of missionary-operated health centers also appeared in the rural scene. The Gondar Public Health College was established in 1954 to train health officers, nurses, and sanitarians to work in rural health centers. By 1961/1962, there were 205 doctors, 344 nurses, and 1,025 health assistants serving in 54 hospitals, 41 health centers, and 354 health stations (Desta and Meche, 1999: 16) (Table 6.1a & b).

In 1966, Ethiopia's first medical school was established in Addis Ababa for the training of Ethiopian physicians and pharmacists. The 1960s and 1970s saw the establishment of additional training centers for health officers, nurses, health assistants, pharmacists, and sanitary officers. In a series of 5–year plans, the government also set goals to expand preventive health services to all parts of the country (Kloos and Zein, 1993: 135). Health facilities continued to grow: by the early 1970s, the number of hospitals grew to 84, health centers to 91, and health posts to 650. The number of health personnel also showed significant growth over the 1960s: there were 374 medical doctors, 213 health officers, 1,162 nurses, and over 4,000 health assistants (Desta and Meche, 1999: 18) (Table 6.1a & b).

**Table 6.1a  Development of Healthcare Facilities, 1947/48 – 2000/2001**

| Facilities | 1947/48 | 1961/62 | 1972/73 | 1977/78 | 1990/91 | 1998/99 | 2000/01 |
|---|---|---|---|---|---|---|---|
| Hospitals | 46 | 54 | 84 | 84 | 89 | 90 | 110 |
| Hospital beds | 3,400 | 5,158 | 8,415 | 8,623 | 12,106 | 11,371 | 10,736 |
| Health Centers | - | 41 | 93 | 110 | 160 | 294 | 382 |
| Health Stations | 92 | 354 | 650 | 1,133 | 2,292 | 3,251 | 2,393 |
| Health Posts | - | - | - | - | - | - | 1,023 |

**Table 6.1b  Development of Health Human Resources, 1947/48 – 2000/2001**

| Type of Health Personnel | 1947/48 | 1961/62 | 1972/73 | 1977/78 | 1990/91 | 1998/99 | 2000/01 |
|---|---|---|---|---|---|---|---|
| Doctors | 106 | 205 | 374 | 605 | 1,658 | 1,415 | 1,366 |
| Health Officers | - | - | 213 | 430 | 94 | N/A | 296 |
| Nurses | 88 | 344 | 1,162 | 3,645 | 3,575 | 4,774 | 7,723 |
| Health Assistants | - | 1,025 | 4,003 | 7,478 | 10,045 | 10,071 | 7,386 |

Source: Asfaw Desta & Hailu Meche (1999), 'Review of the Health Services of Ethiopia' in *The Proceeding of the 10th Annual Scientific Conference* 19-20 October 1999; Addis Ababa; Ministry of Health (2002), *Health of Health Related Indicators,* Addis Ababa: MOA.

These early efforts to promote healthcare targeted only a very small segment of the population. Hospitals in major urban areas received the lion's share of the public health spending. In 1972, for instance, nine out of every ten Birr spent on health went to hospitals. While Addis Ababa and Asmara (then part of Ethiopia) received 64 Birr and 34 Birr per resident, respectively, the rest of the country received only about 3 Birr per person. In 1971-72, health centers received only 2.1 percent of the total capital budget allocated for health services while hospitals and disease eradication programs consumed over two-thirds of the allotment (Kloos and Zein, 1993: 136).

## Health Services Under the Derg Regime

The Derg regime's health policy put greater emphasis on healthcare, 'rural health services, prevention and control of common diseases, self-reliance, and community participation in health activities' (Kloos and Zein, 1993: 136). It embraced the objective of the 1978 Alma-Ata Conference of the World Health Organization declaration to make 'health for all' a reality by the year 2000. In the 1980s, the Gondar Public Health College became a medical school, a third medical school opened in Jimma, and two more urban hospitals opened. The Derg established a hierarchical national healthcare system that consisted of (from the top) teaching hospitals and central referral regional hospitals, district hospitals, health centers, health stations, and health posts to disburse health services (Kloos and Zein, 1993: 137). The latter three in the hierarchy were identified as the approach most likely to promote primary healthcare (PHC) in rural areas. The PHC strategy emphasized preventive healthcare, the promotion of community participation and health education, and the training of health assistants.

From 1974-1989 the number of healthcare facilities in the country grew. Over 5,000 community health agents—which did not exist before 1974—were established. The number of health stations increased from 650 to 2,292, while the number of health centers increased from 93 to 160. The number of hospitals increased from 84 to 89; and hospital beds, from 8,415 to 12,106. Medical personnel also increased: the number of doctors grew by 291 percent (to 1,658); nurses, by 208 percent (to 3, 575); pharmacists and pharmacy technicians, by 382 percent (to 660); laboratory technicians, by 192 percent (to 831); and health assistants, by 243 percent (to 10,045) (Kloos and Zein, 1993: 142-43; Desta and Meche, 1999, Table 6.1b).

Although these achievements may have seemed impressive, the reach of healthcare services still remained limited due to inequities in their distribution and inadequacy of resources allocated to the health sector. For instance, the share of the government budget allotted to health services dropped from 6.1 percent in 1973-74, to 3.5 percent in 1986-87, and to 3.1 percent in 1990-1991. There were reasons for the decline of health expenditure and the general health system failure of the 1980s. Economic crisis in the 1980s, sharp decline in foreign development assistance, and civil war severely undermined social sector spending. The enormous expenditure on the war effort both in Tigray and Eritrea squeezed out spending on health, education, and other public services.

While the number of health centers and health stations significantly increased, the quality of their services suffered due to insufficient personnel and acute shortages of medical equipment and drugs (Kloos and Zein, 1993: 145).

Public health centers covered less than one-fifth of patient drug requirements because of the limitation on heath expenditure. Few of the health posts and health stations that served the rural population provided services for much of the 1980s; and when they did operate, patients often bypassed these lower tier health facilities that lacked basic health supplies and human resources (Shiferaw, 1999). The attrition rate of health agents serving at community level was high. In 1990, for instance, a little over a third of the nearly fourteen thousand community health agents and less than one-half of the twelve and half thousand trained birth attendants were providing healthcare services in their community—due largely to the lack of support from both the health posts as well as the communities (Kloos and Zein, 1993: 141). In general, toward the end of the 1980s the overall healthcare system performance suffered from severe scarcity of healthcare supplies, inadequate health personnel, and poor motivation of doctors and health workers.

Additionally, salaries and operational costs consumed more than four-fifths of the health expenditure while pharmaceuticals received only 16 percent (Larson and Desie, 1994: 94). Moreover, there was consistent urban bias in the distribution of health personnel and expenditures. Addis Ababa received by far the largest share of the health expenditure during this period: 28 percent in 1983-84 and 40.8 percent 1988-89 (Kloos and Zein, 1993: 146). In their study of the health profile of 52 districts (representing 40 percent of the Ethiopian population) between 1988-91, Larson and Desie found that urban districts received three times the number of physicians and four times the number of nurses in rural districts. About 83 percent of the physicians and 63 percent of the nurses were based in hospitals in urban districts. Hospital budgets accounted for 60 percent of the health expenditure. Overall, urban districts received 20 Birr/person while the share of rural districts amounted to only 0.10 Birr/person (Larson and Desie, 1994: 87, 93-94). Hence, in spite of modest increases in the number of both healthcare personnel and facilities, the allocation of health resources continued to favor a few major urban areas at the expense of the nine-tenths of the population that dwelled in villages and small towns.

## Health Services Under the EPRDF

The EPRDF government that replaced the Derg regime in 1991 issued a document in 1995 that laid out its own healthcare strategy for the country. At the outset, the document finds fault with several aspects of the Derg regime's healthcare policy. First, a substantial amount of the Derg regime's recurrent health budget had been spent on salaries and wages, leaving very little for drugs and medical supplies. Second, emphasis had been on curative care rather than preventive health service. Third, the allocation of services reflected a bias in favor of hospitals located in major urban areas, especially in Addis Ababa. Fourth, the misuse of an incoherent cost recovery program damaged the healthcare system.

Fifth, the implementation of capital expenditures rarely exceeded 50 percent. Finally, the health service system was highly centralized, bureaucratic and non-participatory in its management and service delivery (TGE, 1995: 17).

The EPRDF's healthcare strategy promises to make fundamental alterations in the healthcare system of the country. The EPRDF articulates the changes it plans to undertake in its 1995 *Health Sector Strategy* document (Transitional Government of Ethiopia, 1995: 13-20):

*To strengthen the preventive and promotive health service* The health sector strategy promises to place heavy emphasis on preventing the most common infectious diseases and controlling epidemic outbreaks by making healthcare deliveries 'more accessible, affordable, cost-effective, efficient, and sustainable by establishing primary healthcare units with standard facilities and staffing serving a manageable population and equitably distributed throughout the country.' Additionally, the strategy will give special attention to the improvement of maternal and childcare services, including antenatal, perinatal and postnatal care, family planning services, nutrition education, and immunization.

*To promote curative and rehabilitative care within the family home setting* This involves the provision of curative and rehabilitative services at the grassroots level via trained frontline health workers and traditional health practitioners.

*To provide essential drugs and medical supplies for all levels of the health service* This involves major revamping of the procurement, distribution, storage and utilization systems of drugs and medical supplies, and enhancing the country's capacity to produce drugs, vaccines, and medical supplies. In this regard, the health sector strategy promises to encourage the participation of the private sector in the 'procurement, distribution and production of drugs and [medical] supplies.' It also envisions the creation of a favorable environment to garner financial assistance from international donors and NGOs toward these efforts.

*To document and process health information* The plan underscores the need for the collection of accurate and detailed health information in order to deliver healthcare services, allocate health resources equitably and effectively, and monitor and evaluate health progress.

*To reform the organization and management of the health delivery system* This plan promises to promote decentralized and participatory management and delivery of healthcare services. The plan emphasizes the incorporation of preventive and promotive health education in all health institutions: to encourage community participation in immunization and disease control campaigns, health education, environmental health sanitation, and the maintenance of local healthcare facilities. The sector plan also intends to improve inter-sectoral joint efforts in providing potable water, sanitation and waste disposal facilities, environmental hygiene education, disaster management, and care to groups with special health needs.

*To develop and manage health sector human resource* This involves increasing the number of trained health personnel, especially in the lower and middle level health personnel, to deliver primary healthcare; instituting appropriate pay scales and incentive systems for health workers; instituting in-service training and continuing education programs; and putting in place an effective monitoring and supervision system in order to raise the level of skills and performance.

*To enhance health research and development* The major emphasis in this regard is the promotion of research undertakings that focus on health problems germane to the country.

*To increase public funding for health* The health sector strategy promises to allocate adequate financial resources for both recurrent and capital health expenditures and to disburse such resources more equitably.

The government reduced the previous six-tier organizational structure of the healthcare delivery system to four tiers. The primary healthcare unit, with 10 beds and 13 technical and 12 non-technical staffs, serves up to 25,000 people. At this level emphasis is placed on the preventive aspects of healthcare and on the more easily treatable medical conditions. The second level of healthcare facility is the district hospital with 50 beds, 33 technical and 35 non-technical staffs serving 250,000 people. The third level is the zonal hospital with 100 beds, 60 technical and 30 non-technical staffs to serve 1,000,000 people. The highest level is the specialized hospital with 250 beds, 120 technical and 50 non-technical staffs for every 5 million people (MOH, 2002). The former health posts and health stations were folded into primary healthcare units while health centers were redefined as district hospitals. In the mid-1990s, the government launched a 20-year health plan to be carried out over four five-year increments. The first five-year plan (1997/1998-2001/2002) projected to spend over 5 billion Birr of which 55.4 percent was to come out of the government coffer, 42.5 percent from foreign assistance, and 2.1 percent from user service fees. Over the first five-year period, 51.4 percent of the total planned budget was allocated for service delivery, 27.5 percent for the rehabilitation and expansion of health facilities, 4.2 percent for improving pharmaceutical services, 3.1 percent for human resources development, 1.9 percent for improving healthcare management, 1.3 percent for information, education, and communications, 0.6 percent for monitoring and evaluation, and 0.2 percent for healthcare financing (Desta and Meche, 1999: 20-21).

To measure health progress in the country, the first-five year health plan also set specific goals to be achieved in key health determinants by the end of 2001/2002: for example, decreases in the infant mortality rate from 110-128/1000 in 1996/1997 to 90-95/1000 in 2001/2002; a decrease of the maternal mortality rate from between 560 and 850/100,000 to between 450 and 500; a decrease of the population growth rate from 2.9 to 2.5-2.7 percent per year; and an increase in Expanded Program on Immunization (EPI) coverage (DPT3) from 59.3 percent to 70-80 percent; full immunization from 31.7 percent to 60 percent; antenatal service coverage from 30.4 to 50 percent; contraceptive coverage from 9.8 to 15-20 percent; total health service coverage from 48.5 to 50 percent;[3] and life expectancy from 51 to between 55 and 60 years. Other long-term goals to be achieved by 2015 also included a decrease in total fertility rate from 6.1 to 4 children per woman and an increase in institutional birth delivery from 10.2 to 25 percent. Targets that were set in the realm of health facility expansion also included the additions of 216 public healthcare units, 50 district hospitals, 10 zonal hospitals, and 2 specialized hospitals by 2001/2002 (MOH, 2000a; Shiferaw, 1999).

**Major Health Sector Developments**

In order to achieve the aforementioned goals, the government has established new training facilities and upgraded existing ones to train health workers at all levels of profession. The country now offers three levels of health education: schools for higher-level health education, senior training schools, and junior training schools. Three major higher health education institutions—the Addis Ababa University Medical School, Gondar College of Medical Sciences, and the Institute of Health Sciences at Jimma University—train physicians for a duration of six years plus a further period of one year of supervised clinical practice before awarding the degree. These three medical schools produced a combined total of 556 physicians between 1997/1998 and 2000/2001. Between 1998/1999 and 2000/2001, the three institutions also produced a total of 574 health officers, nurses, environmental health officers, lab technicians, and pharmacists with Bachelor of Science degrees and 1,664 clinical, public health, and mid-wife nurses; lab, radiography, and pharmacy technicians; and environmental health officers with diplomas.[4]

The former Alemaya College of Agriculture was upgraded to a university in 1999 that included a school of public health that now produces health officers with Bachelor of Science degrees and public health nurses, lab technicians, and sanitarians with diplomas. Established in the mid-1990s, Debub University also trains health professionals with Bachelor of Science degrees and diplomas. The two institutions together produced 155 health personnel with Bachelor of Science degrees and 797 with diplomas between 1998/1999 and 2000/2001. Six senior training schools (located in Nekemte, Mekele, Asela, Harar, Yirgalem, and Awasa) offer a two-year diploma program to train clinical, midwives, and pharmacy technicians. These senior health schools produced a total of 1,144 health personnel between 1998/1999 and 2000/2001. Thirteen junior training schools (scattered in 14 cities, seven regional states) that run a one-year certificate program trained 3,151 health officers (clinical, public health, and midwives, pharmacy, laboratory, and X-ray technicians, and environmental health officers) between 1998/1999 and 2000/2001.

The last few years have thus seen tangible growth in human resource developments in the health sector. In the years 1997/1998 to 2000/2001, the five institutions of higher learning in the health field annually produced on average 139 doctors, 1,258 clinical and public health nurses, 186 midwives, 125 health officers, 182 environmental health officers, 150 pharmacists and pharmacy technicians, and 301 lab technicians. During the same period, 19 senior and junior health training schools combined produced on average 1,074 lower level health professional annually.

While there has been a steady growth in trained health personnel at all levels of the profession, the number of physicians working in public health facilities has declined from 1,483 in 1996/1997 to 1,366 in 2000/2001. The decline may be due to the fact that increasing numbers of doctors move into private practice or private sector employments in the country or abroad. The number of health assistants also declined—from 10,625 to 7,386 during the same period—perhaps because of a shift in training more health personnel in other specialties. For instance, the number of public, clinical, and midwife personnel increased from 4,114 in 1996/1997 to 7,723 in 2000/2001; health officers, from 30 to 296; and paramedics, from 1,788 to 2,758.

The achievements in health sector human resource developments have undoubtedly been impressive and surpass efforts of the 1980s and the early 1990s. However, many critics, including quite a few from the health profession itself, seriously question the quality of training offered at many of the country's health institutions. They point out the glaring shortage of qualified trainers and teaching materials, including books and journals, laboratory teaching equipment and manuals, and essential teaching consumables. Moreover, the institutions offer short, superficial training that lacks adequate field practice. Hence, critics argue, most existing health training institutions recruit intelligent minds but produce mediocre health professionals.

The government, however, sees things differently: it admits that existing health training institutions are far from being equipped with full-fledged educational resources and a perfect learning environment, but it believes that its health trainees receive sufficient basic health instructions and have the ability to tackle health problems relevant to the country. The government also argues that most Ethiopians suffer from diseases that can easily be treated or prevented by simple measures managed by public health officers and nurses, environmental health officers, and frontline health workers. As the government sees it, the most important preventive measures are health education, particularly that directed towards improving environmental sanitation and modifying personal habits, immunization, and routine examination of the antenatal and under-five clinics. The country's health policy-makers contend that health practitioners with 1-2 years of basic health training can manage these and other crucial health problems that affect the majority of the Ethiopian population. Hence, they advocate the allocation of a significant share of the public health resources towards producing a greater number of low level health professionals in a shorter period of time and establishing as many lower level healthcare facilities as possible to reach as many citizens as possible.

By and large, the current government has designed its health policy to make the most of the limited resources for the reduction of diseases. The government sees preventative medicine as the most efficient path to bring tangible progress in healthcare. It argues that a preventive strategy of healthcare will give a better return in human welfare for the limited resources and medical personnel available in the country. For example, investing scarce resources in measures such as the prevention of measles by immunization or the control of malaria by eradicating malaria-carrying mosquito habitats, providing such preventives as insecticide-treated bednets, and promoting health education costs less than the treatment of patients already attacked by these diseases. Thus, the government sees its vaccination programs as the most cost effective strategy with regard to disease: vaccination halts disease transmissions, reduces morbidity and mortality, and costs less than medical treatment.

This policy is not without critics, however. Some clinical health professionals question the government's emphasis on preventive medicine. First, they argue curative medicine should play a prominent role because at present too many diseases afflict too many people in the country. While preventive medicine plays a big role in reducing the incidence of disease, it cannot comprise a substitute for curative medicine, especially in an impoverished country. Second, many health problems have no feasible preventive solutions; and even where preventive solutions exist, such solutions may not prevent all cases of particular diseases. Third, curative medicine has itself a preventive role to play. For example, providing adequate treatment to patients with contagious diseases reduces the spread of such diseases and thus reduces potential deaths and economic burdens to society. Fourth, clinical health services can generate information from patients concerning existing outbreaks or epidemics in certain localities. Such information can facilitate immediate action against the spread of the disease to the wider community. Regardless of the policy controversy, however, far greater public funds are still apportioned for curative rather than for preventive medicine.

Health personnel trainings for lower level healthcare facilities have shifted towards specialization. Junior nurses have replaced the former health assistant positions that existed under previous administrations. Former health assistants provided comprehensive healthcare services that included treating patients for the most common diseases, providing maternal and childcare services, participating in preventive health services (such as providing sanitary education and immunization), and supervising community health agents and traditional birth attendants. Health assistants were thus involved in both clinical and preventive healthcare practices. By contrast, the current senior and junior nurses are trained for specialized clinical or preventive healthcare services. For example, senior nurses (with a two-year diploma) and junior nurses (with a one-year certificate) specialize in one of several lower level healthcare positions such as clinical nurse, public health nurse, and mid-wife nurse. Laboratory, pharmacy, and X-ray technicians and sanitarians are also trained at senior (diploma) and junior (certificate) levels.

The trend toward specialization in the training of healthcare workers has drawbacks. First, a single-purpose trainee is confined to a narrow specialization and is not allowed to shift to different areas of medical interest. A person trained as a midwife has to remain in that task and cannot design an alternative career path within the profession. Second, a single-purpose trainee may not be able to provide assistance in an emergency situation that requires her/him to perform tasks outside her/his specialization.

The country has placed an increasing emphasis on Frontline Health Workers (formerly called community health agents) to render basic health service at the grassroots level. They are usually selected from the community largely based on their experience in traditional health practices (for example, as a birth attendant), but their training lasts a maximum of six months. The services they render to their communities may include treating minor illness and injury, providing maternal and child health services, providing preventive services (hygiene education, for example), collecting health information, and reporting epidemic outbreaks. They may also refer patients with serious illness to the nearest health facility. The MOH provides them with basic first aid kits, delivery kits, drugs for common diseases, and basic health consumables. About 6,893 Frontline Health Workers served in the countryside in 2000 (MOH, 2000a: 21).

The last decade has also seen significant growth in healthcare facilities, both at lower and higher levels. From 1990/1991 to 2000/2001, the number of hospitals increased from 89 to 110; the number of health centers, from 160 to 382; and the number of health stations and posts, from 2,292 to 3,426 (Table 6.1a). Recent years have also witnessed an increase in the number of private hospitals and clinics in major urban areas, especially in Addis Ababa, partly as a result of the inability of the public health system to provide adequate services and government promotion of private sector involvement in the provision of healthcare services.[5] The number of private clinics increased from 541 in 1996/1997 to 1,170 in 2000/2001. There were 9 privately owned hospitals in 2001, eight of which were located in Addis Ababa. Addis Ababa also had 340 (29 percent) of the country's private clinics (MOH, 2001: 23) (Table 6.2). While the MOH is still the chief healthcare provider in the country, and while only those with enough money can access private healthcare, healthcare within the private sector is becoming increasingly important. There is clear evidence that much of the medical and health facilities and expertise are concentrated in few major urban centers. In 1999, 62 percent of all doctors, 30 percent of hospital beds, and 46 percent of nurses were located in the capital (MEDaC, 1999: 29).

**Table 6.2 Healthcare Facilities by Regional States, 2000/2001**

| Region | Pop. (1000s) | Hospital | | Health Centers | | Health Station | Health Post | Clinics |
|---|---|---|---|---|---|---|---|---|
| | | No. | Beds | No. | Beds | | | |
| Tigray | 3,797 | 13 | 1,217 | 29 | 46 | 179 | 103 | 2 |
| Afar | 1,234 | 2 | 208 | 5 | 28 | 53 | 36 | 2 |
| Amhara | 16,748 | 16 | 1,207 | 77 | 303 | 536 | 361 | 179 |
| Oromiya | 23,023 | 28 | 1,981 | 114 | 0 | 795 | 145 | 435 |
| Somali | 3,797 | 6 | 330 | 10 | 0 | 91 | 0 | 0 |
| B/Gum | 551 | 2 | 171 | 7 | 254 | 71 | 24 | 2 |
| SNNPR | 12,903 | 12 | 963 | 107 | 0 | 443 | 290 | 158 |
| Gambela | 216 | 1 | 89 | 5 | 0 | 38 | 0 | 25 |
| Harari | 166 | 5 | 328 | 2 | 10 | 19 | 7 | 14 |
| Addis Ababa | 2,570 | 23 | 2,126 | 24 | 136 | 161 | 47 | 340 |
| Dire Dawa | 330 | 2 | 245 | 2 | 8 | 7 | 10 | 13 |
| *National* | 65,344 | 110 | 10,736 | 382 | 785 | 2,393 | 1023 | 1,170 |

Source: Ministry of Health (2001), *Health and Health Related Indicators*, Addis Ababa: MOH.

The availability of essential medicines in the market has also improved during the last few years. Prior to 1991, the government monopolized the production, importation, and distribution of pharmaceuticals and medical supplies. Since 1991, about 66 private companies have been granted licenses to import and distribute pharmaceuticals (MOH, 2000b: 28). Of the 311 pharmacies operating in the country in 2001, the private sector operated 271 (87 percent) of them. The private sector also operates 240 of the 249 drug stores in the country. The 1,917 drug vendors serving the rural areas are all private.

The proportion of the national budget spent on health services has increased steadily, from an average of less than 2.3 percent annually in the 1980s to 5.5 percent in the 1990s. Per capita health expenditure increased from less than 2.5 Birr to 8.5 Birr annually during the same period. There have been some improvements as well in the implementations of recurrent health budgets. The average level of expenditure increased from 79 percent in the late 1980s to 94.4

percent between 1992/1993 and 2000/2001. The implementation of the health budget has, however, been quite low and erratic: it averaged 65 percent between 1988/1989 and 1992/1993, went up to 98 percent in 1993/1994, declined to an average of 59 percent between 1994/1995 and 1997/1998, shot up to 214 percent in the following year, and precipitously fell to less than 50 percent in 1999/2000 and 2000/2001. Moreover, the proportion of the national budget allocated to the health sector fell from 6 percent in 1996/1997 to 4.8 in 1999/2000 but rose to 7.3 percent in 2000/2001. There was a somewhat steady increase in capital budget allocation, from 5.9 percent in 1996/1997 to 13.6 percent in 2000/2001 while budget for recurrent expenditure fell from 6 percent to 4 percent during the same period (MOH, 2001: 3). This decline may have been caused by a huge jump in defense expenditure—from $95 million in 1997/1998 to $777 million in 1999/2000 (Bhalla, 2001)—caused by the war with Eritrea.

One can discern patterns in the distribution of health budgets among regional states. In 2000/2001, regional states with a smaller population had a much higher per capita share than those states with a larger population. Among the smaller regional states, Gambela received the largest per capita health budget (45.3 Birr) followed by Harari (44.8 Birr), Addis Ababa and Benishangul-Gumuz (22.0 Birr), Dire Dawa and Afar (18.0 Birr), and Tigray (13.8 Birr). The per capita budget share of the largest states ranged from 4.9 Birr in Amhara to 5.4 Birr in the SNNPR to 5.8 Birr in Oromiya.

## Critical Healthcare Problems

In spite of some genuine efforts to improve the healthcare system over the past ten years, the country still needs to invest far more of its resources in health services. The current expenditure on health is extremely low compared to the health needs of the fast growing population. Moreover, existing health delivery services are deteriorating. Public healthcare institutions are in appalling condition. Inadequately funded and ill equipped, they suffer from shortages of everything—doctors, nurses, lab technicians, drugs and ambulances. Their infrastructure—buildings and grounds—is crumbling for lack of maintenance. Physicians and nurses are grossly underpaid and overworked. Consequently, many doctors have been forced either to leave the public service and join the growing private practices in Addis Ababa and other major urban areas or to emigrate abroad. The emigration of trained doctors and other health professionals is a major problem. For instance, there were 1,658 medical doctors in the public health sector in 1990/1991 (Desta and Meche, 1999), but there were only 1,366 in 2000/2001. There are probably more Ethiopian physicians practicing in the New York metropolitan area than in Addis Ababa itself. According to the International Organization for Migration, Ethiopia tops all African nations in the emigration of medical professionals.[6] More than one-half of the doctors trained in Ethiopia leave the country (Shinn, 2001: 3). The government could alleviate the problem of emigration by adopting policies that encourage health professionals to stay in the country. For instance, the government could establish career structures with an attractive salary scale and a reward system that recognizes professional excellence and creativity, and provide professional development opportunities, adequate resources for infrastructure and research, good working conditions, and sufficient professional independence.

Generally, the quality of healthcare depends not only on the ability of the health institution to provide adequate quality health services but also on the manner in which professional health personnel interact with their patients. Unfortunately, most health personnel in the country, especially those working in public health institutions, have the reputation for being impolite, supercilious, and hostile to their patients. Perhaps overworked and underpaid, most health professionals do not show a humane and genial disposition toward their clients and often give the impression that they dislike their job intensely.

The diagnosis of diseases is a fundamental practice in medicine, and to be able to do that a health establishment requires a laboratory with adequately trained lab technicians, essential outfits of laboratory apparatuses, and lab manuals. One of the glaring weaknesses of the present public healthcare system, especially at the lowest tier, is the difficulty their staffs have in making a definite diagnosis of the most common diseases because of the partial, or in many cases total, lack of critical laboratory tools. For example, in their study of laboratory services in 27 health centers in the Amhara  Regional State, Kassu and Aseffa (1999: 239-42) found that nearly nine out of ten health centers did not perform the Weil-Felix agglutination test for typhus fever, the following tests: Widal agglutination test for typhoid fever, the VDRL (venereal disease research laboratory) or the RPR (rapid plasma reagent) test for syphilis; the cross matching test for blood compatibility (a necessary test for blood transfusion); the blood grouping test for determining blood type (a very critical test for pregnant mothers and blood transfusions); the stool concentration test for intestinal parasites; the white blood cell count (a necessary test for evaluating infection for almost all patients); the KOH (potassium hydroxide) test for fungus infections; or the CSF (cerebro-spinal fluid) microscopy test for meningitis; all due to lack of necessary lab equipment. Over 80 percent of the labs had no refrigerators, about 90 percent had no centrifuges and sterilizers, and between 50 and 90 percent had no reagents and chemicals critical for many lab tests. Because of the lack of adequate diagnostic facilities and laboratories, doctors and healthcare officers must engage in symptom-based presumption to help patients. Unfortunately, this kind of practice often results in misdiagnosis and hence increases the risk of prescribing inappropriate medication.

*Reproductive Health Services* According to the 2000 *Demographic and Health Survey*, only 8 percent of Ethiopian women 15-44 years of age had access to contraceptives. The survey found that 61 percent of urban married women and 42 percent of rural married women expressed their desire for family planning services. The Survey estimated contraceptive use would have jumped to 44 percent if family planning services were available (CSA, 2001a: 90-91). Moreover, the current family planning services are overwhelmingly geared toward the provision of contraceptives—hormonal pills, diaphragms, and condoms—to married women with children. Unfortunately, contraceptive services alone are inadequate for safeguarding women's reproductive health. Contraceptive prevalence rates do not indicate how far family planning programs have improved women's reproductive health vis-à-vis the problem of unwanted pregnancy, infertility, reproductive tract infections, and sexually transmitted pathogens.

Abortions kill many young women, especially in urban areas. Abortion was the second leading cause of hospital mortality in 1994/1995 (MOH, 1998a). This is due largely to the lack of contraceptives and other reproductive health services for young women and the absence of safe abortion services. Because abortion is illegal in the country many young women with unplanned pregnancies are forced to seek abortions performed under dangerous life-threatening conditions. According to the WHO, 'more Ethiopian women die in hospitals from illegal abortion complications than for almost any other medical reason' except tuberculosis.[7] It is estimated that seven in ten women who are brought to the hospital suffering from serious problems after back street abortions die. Most of these deaths occur among women between the ages of 16 and 20. Existing family planning programs are often designed not for young single women, but for married women in an attempt to limit family size. Even when young women manage to access family planning services, many shy away from such services because they worry about divulging their sexual activity and because they fear the hostile or judgmental attitudes of family health service providers.

In developing countries one woman out of 65 dies of reproductive health problems. In developed countries, by comparison, one woman out of 2,125 dies from similar causes. According to the WHO, in the developing countries as many as 1,600 women die from the complications of pregnancy and childbirth every day, and almost nine-tenths of these deaths occur in Sub-Saharan Africa and Asia (WHO/UNICEF, 1996). Ethiopia is among the ten countries of Africa (the others being Angola, Chad, Central African Republic, Democratic Republic of Congo, Lesotho, Liberia, Mali, Niger, and Sierra Leone) with the highest maternity mortality (Panafrican News Agency, 2001). There were 871 maternal deaths for every 100,000 live births during the period 1994-2000, more than twice the world average (430) (UNICEF, 1997: 48). An Ethiopian woman faces a 1 in 9 lifetime risk of dying from complications associated with pregnancy, delivery, or perinatal infection. By comparison, 'over a woman's lifetime, the risk of dying from pregnancy-related causes is about 1 in 16 in Africa, 1 in 65 in Asia, 1 in 130 in Latin America, 1 in 6,000 in the United Sates, and 1 in 10,000 in Europe' (Speidel, 2002: 90).[8] The major causes of maternal death in Ethiopia are abortion (54 percent), hemorrhage and hypertensive pregnancy diseases (11.5 percent), obstructed labor, mainly ruptured uterus (8.3 percent), anemia (4.8 percent), indirect obstetrical causes, such as infectious viral hepatitis (3.3), and others (23.8 percent) (FDRE, 1995: 15-16). Maternal deaths account for one-quarter of all deaths of women 15-49 years of age (CSA, 2001a: 109). For each woman who dies, many others experience lifetime ailments or injuries that significantly reduce their health productivity.

Maternal deaths occur because women have little or no access to basic medical care during pregnancy. Antenatal care is extremely vital because it can help healthcare providers to 'detect and manage existing diseases, recognize and treat complications early, provide information and counseling on signs and symptoms of problems, recommend where to seek treatment if complications arise, and help women and their families prepare for childbirth' (WHO/UNICEF, 1996). Unfortunately, very few women in Ethiopia access antenatal service owing to the scarcity of obstetric care. Only 34.7 percent of women in Ethiopia received antenatal care services during pregnancy in 2000/2001 (Table 6.3). The distribution of such services ranged between 3.2 percent in Somali State and 13.3 percent in Afar State to 45.5 percent in Dire Dawa City, 49.5 percent in Harari State, 79 percent in Tigray State and 92.3 percent in Addis Ababa. Only one-third or less women in the most populous regional states such as Oromiya, Amhara, and the SNNPR had access to antenatal services. Because public medical services are located overwhelmingly in urban settings, urban women receive higher rates of maternal care than their rural counterparts. While over two-thirds of urban mothers receive antenatal care, the corresponding figure for pregnant rural women is only 22 percent (CSA, 2001a: 111).

The Central Statistical Authority's national health survey indicates that the level of the mothers' education is highly associated with the level of antenatal care services accessed. Seventy-two percent of mothers with a secondary education or higher accessed such services compared to 45 percent of mothers with some education and only 21 percent of mothers with no education at all (CSA, 2001a: 112).

The presence of a healthcare facility that provides antenatal care does not guarantee the full use of the service available. Of the 12 or so antenatal visits recommended by health professionals, only one in ten mothers made four or more visits during her pregnancy. The median number of visits for mothers who accessed antenatal care services was fewer than three times. Mothers also tended to make their first visit to antenatal care well into their pregnancy (CSA, 2001a: 113). Cost of transportation and low level of awareness of the service are important reasons for the low number of antenatal visits. There is, however, another factor women take into account besides these. This is the value of time spent waiting at a healthcare facility and traveling to and from it. Based on a sample survey of 34 health facilities in the country in 1998, the average waiting time (that is, the average time between the time a patient arrives at a health facility and the time the patient is seen by a health personnel) at rural primary care

**Table 6.3 Maternal Health Service Achievement and Coverage by Regional States, 2000/2001**

| Region | Antenatal (% coverage) | Attended Delivery (% coverage) | Postnatal Services (% coverage) | TT2 + Achievement & Coverage (% coverage) |
|--------|------------------------|--------------------------------|---------------------------------|-------------------------------------------|
| Tigray | 79.0 | 41.0 | 37.9 | 33.8 |
| Afar | 13.3 | 3.6 | 0.6 | 2.1 |
| Amhara | 30.1 | 7.9 | 6.1 | 36.1 |
| Oromiya | 33.0 | 6.3 | 5.5 | 29.5 |
| Somali | 3.2 | 2.0 | 0.2 | 2.9 |
| B/Gum | 31.7 | 14.2 | 2.2 | 6.6 |
| SNNPR | 34.0 | 9.0 | 2.7 | 25.7 |
| Gambela | 38.0 | 12.8 | 4.8 | 24.9 |
| Harari | 49.5 | 21.6 | 11.2 | 44.4 |
| Addis Ababa | 92.3 | 37.4 | 14.3 | 56.7 |
| Dire Dawa | 45.5 | 22.5 | - | 52.8 |
| *National* | 34.7 | 10.0 | 6.8 | 29.3 |

Source: Ministry of Health (2001), *Health and Health Related Indicators*, Addis Ababa: MOH, pp. 16-17.

facilities was 4 hours and 40 minutes in rural areas and 6 hours and 44 minutes at urban hospitals (Policy and Human Resource Development Project, 1998). Overall, studies elsewhere in other developing nations like Ethiopia show that there are a number of reasons for the inability of women to make use of antenatal care, even when it is available, including: lack of freedom to attend antenatal care; lack of education on the importance of antenatal care; the cost of care; problems in the real or perceived quality of care; long distance to healthcare facilities in poor transportation over poor roads; traditional beliefs and practices regarding pregnancy and delivery; transport and other costs (women often have to take their younger children to visit a health facility, requiring additional transportation cost and provision of food); and poor quality of care (for example, poor relationships between healthcare staff and their patients, long waits, administrative red tape, and the absence of privacy and emotional support) (Wallace, 1995:17). The fact that health workers are predominantly male may also limit the use of reproductive health services by women.

The lack of delivery care poses another major problem for expectant mothers. Delivery under the supervision and care of health personnel reduces the health risks of mothers and the newborns. A trained birth attendant can provide a safe, hygienic delivery and deal effectively with complications (WHO, 2001a: 3). Unfortunately, most deliveries in Ethiopia take place in situations where the woman either has no help or is cared for by family members or untrained traditional birth attendants. In 2000, only one in ten deliveries in the country took place with the assistance of a trained birth attendant, a doctor, a nurse, or a trained midwife. Relatives or some other persons assisted 58 percent of the deliveries; untrained traditional birth attendants assisted 26.4 percent of the deliveries; and 5.7 percent delivered without any assistance (CSA, 2001a: 119). Regional and urban versus rural discrepancies existed in the provision of delivery care. On the short end of the service were Somali (2 percent), Afar (3.6 percent), Oromiya (6.3 percent), and Amhara 7.9 percent), while on the high end were Harari (21.6 percent), Dire Dawa (22.5 percent), Addis Ababa (37.4), and Tigray (41 percent) (MOH, 2001: 5). A health professional or trained traditional birth attendant assisted less than 6 percent of the deliveries in rural areas compared to 45 percent in urban areas (CSA, 2001a: 119). For cultural reasons, untrained traditional birth attendants are still the most popular providers of delivery services in rural places and in small urban areas.

The most deficient of all maternal services is postpartum care. Care during the postpartum period is extremely important because according to the WHO most maternal deaths occur during the early postpartum period. Postpartum care gives healthcare personnel an opportunity to monitor breastfeeding and handle any troubles that might arise (WHO, 2001a: 3). In Ethiopia less than 7 percent of mothers receive postpartum care. Even in Addis Ababa, where there are by far more per capita healthcare services than any other region in the nation, only 14 percent of mothers accessed postpartum care. The service is almost absent in Somali and Afar regions, and in heavily populated regions it reaches only a very small fraction of mothers: SNNPR, 2.7 percent; Oromiya, 5.5 percent; and Amhara, 6.1 percent. With about 38 percent postpartum service access, Tigray mothers are by far the most fortunate (Table 6.3).

Tetanus toxoid vaccination during pregnancy is far from widely available. Tetanus toxoid vaccination during pregnancy is very essential to prevent neonatal tetanus, which kills many newborns. Infants account for almost two-thirds of all tetanus fatalities (Lidetu and Oqubaghzi, 1993: 192). Only 29 percent of pregnant women were immunized with two or more doses of tetanus toxin during pregnancy in 2000. Conspicuous rural-urban and regional discrepancies exist in the delivery of tetanus toxoid vaccination. Urban mothers are three times more likely to get vaccinated than rural mothers. Vaccination rates were extremely low in Afar (2.1 percent), Somali (2.9 percent), and Benishangul-Gumuz (6.6 percent). Coverage levels in the rest of the regions ranged from 25.7 percent in Gambela to 56.7 percent in Addis Ababa (Table 6.3).

There are many factors that are responsible for the low utilization rate for maternity health services. The most obvious factor is the lack of healthcare facilities. Other factors may include: 'distance from health services; costs, including the direct fees as well as the cost of transportation, drugs and supplies; multiple demands on women's time; and women's lack of decision making power within the family' (WHO, 2001a: 3). Studies have also shown that women tend to be reluctant to use services that are deficient in quality and where healthcare providers treat them poorly (Zahr, 1997).[9]

Infant and under-five mortality rates in Ethiopia are among the highest in the world. Infant mortality was 116/1,000 live births in 2001, down from 143/1,000 in 1980 (World Bank, 2003: 112). The under-five mortality rate per 1,000 liver births declined from 213 in 1980 to 193 in 1990 to 172 in 2001 (World Bank, 2001b: 276 & 2003: 18). There were an estimated half a million under-five deaths in 1995 (Amha, et. al., 1995): this figure constitutes nearly 50 percent of all deaths in the country (Gebre, et. al., 1999: 101). In the early 1990s, diarrhea and immunizable diseases (such as measles, pertussis, neonatal tetanus, poliomyelitis, tuberculosis, and diphtheria) were killing an estimated one-third of a million children every year (Lidetu and Oqubaghzi, 1993: 191). In 2001, only one in five children at 12 months of age was fully immunized.[10] Only Tigray State (58 percent) and Addis Ababa (49.8 percent) reached half of the target population of children (Table 6.4). There were no noticeable changes in immunization coverage over the last 4 to 5 years; coverage stayed at between 38 and 43 percent.

**Table 6.4 Immunization Coverage by Regional States, 2000/2001**

| Region | Target Population | BCG (% achieved) | DPT$_3$ (% achieved) | Measles (% achieved) | Fully Immunized (% achieved) |
|--------|-------------------|------------------|----------------------|----------------------|------------------------------|
| Tigray | 150,696 | 86.0 | 78.7 | 71.3 | 58.00 |
| Afar | 41,528 | 4.5 | 2.6 | 4.2 | 1.7 |
| Amhara | 657,664 | 73.0 | 55.4 | 51.4 | 26.9 |
| Oromiya | 960,574 | 54.6 | 39.1 | 31.6 | 25.6 |
| Somali | 126,241 | 8.7 | 5.6 | 5.5 | NA |
| B/Gum | 21,119 | 31.1 | 19.3 | 19.3 | 9.5 |
| SNNPR | 559,863 | 45.3 | 30.8 | 26.4 | NA |
| Gambela | 7,605 | 52.3 | 39.9 | 32.5 | NA |
| Harari | 6,354 | 79.1 | 52.5 | 40.0 | 38.5 |
| Addis Ababa | 67,241 | 58.8 | 56.5 | 49.8 | 49.8 |
| Dire Dawa | 13,412 | 54.4 | 40.4 | 32.1 | NA |
| *National* | 2,598,855 | 56.0 | 41.9 | 36.5 | 21.1 |

Source: Ministry of Health (2001), *Health and Health Related Indicators*, Addis Ababa: MOH.

The only exception is the polio vaccination program that has been the most successful in reaching 78 percent of rural children and 94 percent of urban children (CSA, 1999: 45).[11]

Even when immunization service is available, not all children are fully vaccinated. Challi Jira's study of 376 vaccinated children in Jimma town indicated that more than half (53.5 percent) failed to complete their vaccinations. Nearly half the parents who defaulted said they missed the scheduled appointments; more than a quarter said they had no time to make the trip to the vaccination site; and another quarter said their child was sick on the appointment day. Jira found the lower the education and income of parents the higher the default rate (Jira, 1999: 97).

In poor countries like Ethiopia, perinatal mortality (the death of the fetus after 22 weeks of gestation or death of the newborn through the first week of birth) and neonatal mortality (the death of a live-born infant in the first 28 days of life) (Black, 1999: 4) are quite common. It is estimated that globally 5 million children under 1 month of age die each year and 3.4 million of these neonatal deaths occur in the first week of life (WHO, 1996). Infectious diseases (such as acute respiratory infections, neonatal tetanus and sepsis, diarrhea and meningitis) are linked with 30-40 percent of all neonatal deaths or between 1.5 and 2 million deaths annually. Of the nearly three-quarters of a million neonatal deaths associated with acute respiratory infections each year, most are linked to pneumonia, bronchiolitis, or laryngotracheitis (Black 1999: 9).

Based on the findings of a team of public health experts presented at an international conference in Baltimore, Maryland in 1999, Dr. Robert E. Black (1999: 4) of Johns Hopkins School of Public Health summarizes the causes of perinatal and neonatal death in developing countries:

'Most perinatal and neonatal deaths are caused by infectious diseases, such as sepsis and pneumonia; pregnancy-related complications, such as placenta previa and abruptio placentae; and delivery-related complications, including premature birth, intrapartum asphyxia and birth trauma. Additionally, there are many indirect causes of early infant death, including poor maternal health, untreated maternal infections, including sexually transmitted diseases, urinary tract infections, and chorioamnionitis. Failure to fully immunize adolescent girls and pregnant women also increases neonatal deaths from tetanus, as does unsanitary delivery and umbilical cord care. Premature birth, fetal malnutrition, and failure to exclusively breastfeed also contribute to the risk of early death. Other indirect causes of prenatal and neonatal death include inability to recognize severe illness in a newborn, poor care-seeking behavior, and inadequate access to good quality care. Underlying these direct and indirect causes is widespread poverty, illiteracy and gender discrimination faced by both mother and her female children in developing countries.'

In Ethiopia, perinatal mortality rate was 52 per 1,000 live births in 2000 (CSA, 2001a: 104). This is largely attributable to extremely poor maternal healthcare services, low rates of immunization, and a vast amount of deliveries occurring at home unattended by medical personnel. Maternal malnutrition before or during pregnancy increases the risk of perinatal and neonatal death. Cultural taboos against eating protein-rich foods during pregnancy exacerbate maternal malnutrition. The 2000 national health survey found 36 percent of women stopped consuming cheese and butter during pregnancy, 29 percent avoided vegetables, 27 percent avoided milk, 15 percent avoided meat, and 12 percent avoided fruits (CSA, 2001a: 104). Iron deficiency anemia—which is not considered a major public health problem in Ethiopia—is prevalent among infants and children as well as pregnant and lactating women. Nearly one in five women age 15-59 is anemic. An estimated 42 percent of pregnant women between 1975 and 1991 were considered anemic (World Bank, 2000b: 193). Low protein-energy intake and poor weight gain during pregnancy cause low birth weight, a major contributor to neonatal morbidity. The average weight gain during pregnancy for women in developing countries is 12 kilograms; in Ethiopia it is only 5.6 kilograms (FDRE, 1995, 15-16). A study by the UNICEF estimates that nearly one in five babies born in Ethiopia weighs less than the standard for a healthy-sized baby—2.5 kilograms—largely because of maternal malnutrition during pregnancy. This estimate cannot be accurate, however, because most of the data on underweight births comes from hospital records, thus leaving out the vast number of babies delivered at home (UNICEF, 1997: 20). Maternal malnutrition may increase the likelihood of maternal infection and hamper the formation of the fetal immune system (Meade and Erickson, 2000: 16). Malaria, which is widespread in lower elevations of Ethiopia, may cause up to 30 percent of low birth weight (Black, 1999: 3).

Malnutrition is one of Ethiopia's most serious threats to health. National and small scale surveys all show that malnutrition is prevalent across gender and age groups, though women and children are the most malnourished segment of the population. Malnutrition in Ethiopia results from the under-consumption of protein and energy deficiencies in important micronutrients such as iodine, vitamins A, B1, C, and D, and iron (UNDP, 1998: 17; FDRE, 1995: 30-32). Nutritional deficiencies are a major cause of morbidity and mortality. The average daily per capita intake of energy is estimated at 1,750 calories, about 20 percent below the daily average requirement (TGE, 1995). Malnutrition is particularly high among children: 48 percent of children under age 5 suffer from malnutrition, the highest in Africa (World Bank, 2001b: 276). According to the 1998 CSA *Welfare Monitoring Survey*, nearly 55 percent of the all children 3 to 59 months suffered from chronic malnutrition (CSA, 1999: 129).[12] Nearly three-fifths of all childhood deaths are attributed to protein-energy malnutrition (USAID, 2001: 5). The 2000 CSA Demographic and Health Survey also found that 52 percent of children under age 5 years were stunted (short for their age), 26 percent were severely stunted, and 11 percent were wasted (thin for their height). The proportion of wasted children was highest in the 12-24 months age group indicating insufficient food supplementation to be the cause. Children in rural areas were more than twice as likely to be wasted than those living in urban areas. Forty seven percent of the children were also found to be underweight and 16 percent were severely underweight. Children below the age of six months suffered much lower rates (11 percent or less) of stunting and underweight due largely to breastfeeding (CSA, 2001a: 155). The percentage of infants exclusively breastfed at 4 months was 48.5 in 2000; at 6 months, it was 38.1 (CSA, 2001a: 143). According to the UNICEF, low birth weight increases a child's chances of developing blindness, deafness, epilepsy, cerebral palsy, and mental retardation between the ages of 1 and 3. Stunting diminishes the function of the immune system and limits the learning capacity and performance of the child. Stunting also increases the likelihood of maternal mortality, for it constrains the growth of the girl's pelvis and thus creates the risk of obstructed labor (UNICEF, 1997: 20-21).

The consequences of an inadequate diet are more damaging to children than to adults because children need food for growing. Children deprived of adequate nutrients may suffer from delayed maturation, permanent brain damage, or even death. One of the most common diseases of dietary insufficiency in children one to three years of age is *kwashiorkor*. Children suffering from this disease are severely deprived of protein and may display 'diarrhea, anemia, reddening of the hair, edema, ascites (distended belly), fatty degeneration of the liver, failure to grow, and apathy with extreme weakness' (Moore, 1999: 4.12). *Marasmus* is another common nutritional disease in children one year of age who 'lack sufficient food and show signs of wasting, failure to grow, and complications of infections with multiple deficiencies' (Moore, 1999: 4.12).[13]

*Traditional practices and women's health*  Ethiopian women 'suffer the double deprivation of low overall achievement in human development' (as measured by income, literacy rate, and enrollment into primary, secondary, and tertiary education). Women have less food, less education, fewer employment opportunities, less healthcare, less access to the means of livelihood, and less personal protection. Ethiopia ranks third from last among 143 countries of the world regarding gender-related development. It also ranks fourth from last among 174 countries regarding gender empowerment (as measured by parliamentary seats and administrative and managerial positions held by women, proportion of women in professional and technical works, and women's share of the gross domestic product) (UNDP, 2000: 168).

Women in particular are exposed to greater health risks in the course of securing the necessities of daily life. Women are more exposed to environmental health dangers in and around the house because they bear heavier domestic responsibilities. Women's health and food needs are most often inadequately met. Yet, women bear the responsibility for preparing and cooking food in the household. While women are often responsible for family welfare, they have little or no access to the means of livelihood. In rural areas, women toil on the fields, lug food, firewood, and water, and rear the young children. Citing a World Bank Report, the National Program of Action for Children and Women (Ethiopia) describes a 'typical' Ethiopian woman in the following way (FDRE, 1995: 8-9):

'The typical Ethiopian woman is impoverished, living on the brink of subsistence, mostly in rural areas. She is given in marriage early, bears many children without the benefit of family planning advice and lacks education or information about hygiene and nutrition. She lives in conditions of very high maternal and infant mortality, and child mortality. Lack of education, high fertility, infant mortality and morbidity, and generally poor health conditions are obvious inter-related factors constraining women's productivity and welfare. Moreover, especially in rural areas, she performs laborious household chores—including toting fuel-wood and water over extended distances—which sap much of her energies for more productive work in farming. In urban areas, she is mostly engaged in low-income jobs. She has little time or energy to devote to contemplating how to improve her well-being and economic productivity. She is a victim of her situation, without the capacity to initiate change within the quagmire of her poverty, high fertility, poor health and domestic drudgery.'

As if such deprivation were not burdensome enough, Ethiopian women continue to suffer from many harmful traditional practices that are deeply rooted in antiquated cultural and religious rituals—that wound the bodies and minds of many girls and women. The National Committee on Traditional Practices in Ethiopia (NCTPE) 1997 survey among 65 ethnic groups in Ethiopia found more than 140 types of harmful traditional practices (HTPs). Many of these HTPs identified by the survey affect the health of women: for example, 72 percent of the ethnic groups practiced female genital mutilation (FGM), 73 percent massage the abdomen of a pregnant woman, 41 percent shake a pregnant woman during a prolonged labor, 71 percent apply a crude procedure to hasten the placenta, 48 percent induce bleeding after the expulsion of the placenta, 55 percent force early marriage on young girls, 69 percent sanction abduction, and 48 percent restrict women from consuming certain types of food (NCTPE, 1999: 4).

Of all the traditional practices, FGM is the most detrimental to women's reproductive health. FGM is routinely practiced in both Muslim and Christian communities. Between 80-85 percent of women in the country undergo FGM. Almost all women in the Afar and Somali Regional States, 90-95 percent of women in Dire Dawa and Oromiya, 80 percent in Amhara and Addis Ababa, 74 percent in Benishangul-Gumuz and SNNPR, 43 percent in Gambela, and 36 percent in Tigray have undergone FGM. About 92 percent of FGMs were performed by a traditional practitioner; about 6 percent, by a traditional birth attendant; and about 1 percent, by a trained health officer (CSA, 2001a: 33-35). FGM has widespread support among Ethiopian women.[14] The WHO classifies FGM into Type I (involving 'excision of the prepuce, with or without excision of part or all the clitoris'), Type II (involving 'excision of the clitoris with partial or total excision of the labia minora'), Type III (involving 'excision of part or all of the external genitalia and stitching/or narrowing the vagina opening—infibulation'), and Type IV (involving 'pricking, piercing or incising of the clitoris and/or labia; stretching of the clitoris and/or labia; cauterization by burning of the clitoris and/or surrounding tissue; scraping of tissue surrounding the vagina orifice (angurya cuts) or cutting of the vagina (gishiri cuts); introduction of corrosive substances or herbs into the vagina to cause bleeding and to tighten or narrow the vagina; any other procedure that falls under the definition of FGM given above') (WHO, 1998: 6). In Ethiopia, Type I and II are common except in the Somali Regional State, where type III is practiced. All the different types of FGM have serious consequences on women's health.[15]

There are other traditional practices that are harmful to women. Early forced marriage (between ages 10-15) for women takes place in many parts of the country, especially in rural areas. A 1997 survey found that 62 percent of women in the Amhara Regional State, 53 percent in Tigray, 50 percent in Benishangul-Gumuz, and 51 percent in Addis Ababa were married below the age of 15 (NCTPE, 1999: 35). Early marriage is usually performed without the will of the female child. The age difference in such forced marriage is so big that the female child experiences physical impairment and is prone to contracting sexually transmitted diseases like AIDS. Early pregnancy also increases the risk of maternal mortality because of obstructed labor stemming from the immaturity of the reproductive system. Even when a child-mother survives such an ordeal, sustained pressure of the fetus on the birth canal may cause fistulae (holes in the birth canal that allow leakage of urine or feces from the bladder or rectum) (NCTPE, 1999: 41), leaving her with permanent damage and no social, family, or marital life. The exact number of fistula sufferers in Ethiopia is unknown, 'but each year 8,500 Ethiopian women develop new fistulas' (Kristof, 2003).[16]

Marriage by abduction is another form of violence against young women and is a common practice in rural and small urban areas of Afar, Oromiya, and SNNPR, and to a lesser degree in Addis Ababa, Amhara, Somali, and Harari. This happens mostly when the parent and/or the young woman turns down a proposed marriage or the would-be husband is unwilling to give a dowry or pay for a wedding or both (NCTPE, 1999: 48). This abominable practice almost invariably results in the physical and psychological abuse of the kidnapped woman. It may also lead to her suicide.

Rape is one of the most severe widespread forms of violence perpetrated against women in Ethiopia: few Ethiopian men find it an infringement of women's rights. Men perpetuate the old false belief that a woman's 'no' constitutes a 'yes' and that 'women enjoy forced sex.' Rape is rarely considered the man's fault; rather women are often wrongly blamed for provoking or forcing men. As a result, few women are willing to report rape to the police: women fear social humiliation and believe that the law will not protect them from reprisals.

Ethiopian society maintains an extremely low regard for women. As an Ethiopian women's rights advocate in Addis Ababa put it: 'I was brought up in a strong cultural setting and tradition where women have to work twice as hard to be thought of half as well as men' (Rupphael, 2000). Domestic violence against women is tragically frequent. Cases of husbands amputating women's hands, cutting off breasts, scooping out eyes, knocking off teeth, disfiguring faces, and burning bodies beyond recognition are common gruesome stories written in various newspapers or told on the radio and television or reported by local women's organizations such as the Ethiopian Women Lawyers' Association

(EWLA). Men who commit such criminal acts against their wives, girlfriends, or lovers are routinely set free with little or no jail terms served. Neither the federal government nor regional governments are doing anything to stop this gross human rights violation. The police and the courts are totally insensitive to women's pain and suffering.[17]

In April 2002, the Ethiopian Women Lawyers' Association and other NGOs called on the House of Peoples' Representative to incorporate provisions in the country's amended criminal code that prohibit all forms of violence against women. The call included the legalization of abortion, access to safe abortion and reproductive health services, a ban on female genital mutilation and abduction, and tough provisions against domestic violence. Additionally, these organizations believe that addressing the civil and democratic rights of women is the key to fighting poverty (*Business Review*, 2002: 4). It may be argued that communities have the right to maintain certain traditional practices. However, to allow the exercise of traditional rights does not mean to encourage all traditional practices at any cost. Some traditional practices—especially those that are oppressive and injurious to women—have to be done away with in order for social justice to prevail in the country. While society may appreciate the desirability of maintaining certain traditional practices, it should nonetheless make sure that such practices are not detrimental to some segments of the population.

*The HIV/AIDS challenge*  The spread of Acquired Immune Deficiency Syndrome (AIDS) has added a new dimension of poor health to many people in Ethiopia, especially in urban areas. Like much of sub-Saharan Africa, AIDS has hit Ethiopia hard, spreading like a wildfire: it comprises the leading cause of adult morbidity and mortality in the country. The number of AIDS infections in the country is so staggering it seems implausible. One in every eleven people between the ages of 15-49 years has the Human Immunodeficiency Virus (HIV), the virus that causes AIDS (MOH, 2001: 9), up from one in thirty-one adults in 1993 (MOH, 1998b). In urban areas, one out of every six adults is infected with HIV, up from one in twenty urban adults in 1990 (Khodakevich and Zewdie, 1993: 334). More than 700 people die every day of AIDS. An estimated 1.7 million people have died of AIDS since the first case of HIV infection appeared in 1984 and the first AIDS case was diagnosed in 1986 (MOH, 2000c). The disease has created a generation of orphans. One in twelve AIDS orphans in Africa lives in Ethiopia; that is about 1.2 million orphaned children, according to UNAIDS estimates (MOH, 2001: 41).[18] About 250,000 children under the age of five live with the disease: the infection is passed to the vast majority of these children from the mother during pregnancy, delivery, and breastfeeding (*Daily Monitor*, 2002). In all, there were roughly three million people from a population of 65 million infected with HIV in 2001,[19] up from a quarter of a million in 1990. Nearly three-fifths of those infected were women.[20] Ethiopia has the third largest HIV-positive population in the world, next to South Africa with nearly 5 million people infected with HIV and India with 4 million (Waldman, 2000).

With increased rural-to-urban and urban-to-rural population movement, the infection is also spreading from urban to rural areas.[21] Individuals with large numbers of sex partners, women engaged in sex work, transport workers, soldiers, and migrants separated from their spouses have the highest risk of HIV infections.[22] HIV infection rates for sex workers, for example, increased from 35.9 percent in 1988 to 69.4 in Bahr Dar, from 19.8 percent to 65 percent in Adama, from 5.3 percent to 36.8 percent in Mettu, and from 24.7 percent in 1988 to 73.4 percent in 1999 in Addis Ababa (MOH, 2000c: 2-8).[23] Sexually transmitted diseases (STDs)—syphilis, gonorrhea, lymphogranuloma venereum, and chancroid—are the most prevalent diseases in urban areas, and they tend to increase the risk of acquiring HIV. Over 80 percent of all cases of STDs are transmitted by prostitution (Kloos and Tesfa-Yohannes, 1993: 296), and studies in two communities in Addis Ababa found 30–40 percent of patients with STDs infected with HIV—three times the rate of infection found among the general urban population (MOH, 2000c: 9).

Back in the 1980s, the country's leaders failed to seize the initiative to begin prevention campaigns soon after HIV infection was first reported in 1984 and risk behavior was confirmed. Early public intervention could have promoted safe behavior before the virus became well established. Public plans to tackle the disease were only drawn up in the late 1990s.

Urban-based studies show that in Ethiopia HIV spreads predominantly through unprotected heterosexual contact (intercourse without a condom), accounting for about 88 percent; from an infected mother to her child during pregnancy, childbirth, or through breastfeeding; through transfusions of unscreened blood; and through contaminated needles (UNAIDS, 2000). Nine out of ten reported AIDS cases occur among adults between the ages of 15 and 49, but the peak ages are between 20 and 35 years of age (MOH, 2000c). Premarital sexual intercourse is a natural part of adult lives for most young men and women. Myths about the disease abound, and unsafe sex continues to be rampant. Poverty may also facilitate the spread of the AIDS virus. People who live in abject poverty may not think twice about the risk posed by the HIV. Poverty forces some people to engage in livelihoods that put them in greater risk of contracting HIV/AIDS, for example women who provide sex in exchange for money as a survival strategy. Poverty contributes to weakened resistance and greater vulnerability to infection. There is also a stigma regarding the use of condoms for sex between married people. Wives have little or no choice over whether to use condoms for sex: the mere suggestion risks the husband's irritation or suspicion. Condoms are still seen as tools for sex workers, even though sex workers themselves have a hard time convincing their customers to wear condoms. One out of seven married women in the country is in a polygynous relationship (CSA, 2001a: 76).[24] As Joan Holmes, President of the Hunger Project, puts it: 'there is a direct correlation between women's low status, the violation of their human rights and HIV transmission. Gender inequality has always been dehumanizing. Now it is fatal' (Holmes, 2001).

HIV infections among pregnant women have been on the rise. Surveillance-based studies in a number of major urban sites shows that HIV infection among pregnant women attending antenatal clinics increased from 4.6 percent in 1989 to 15.1 percent in 1999/2000 in Addis Ababa, from 13.0 percent in 1992/1993 to 20.8 percent in 1999/2000 in Bahr Dar, and from 12.3 percent in 1992/93 to 13.6 in 1999/2000 in Dire Dawa. In 1999/2000, 11.5 percent of pregnant women tested HIV positive in Awasa, 11.8 percent in Dilla, 10.7 percent in Soddo, 14.3 percent in Shashamene, and 19 percent in Gambela (MOH, 2000c: 15). An HIV positive mother may pass the virus to her baby before, during, or after childbirth. HIV infects one-third of babies born to HIV-positive mothers (ISAPSO, 2001: 8).

Young women are particularly more vulnerable than men to HIV and other STDs for several reasons. First, the male donates more body fluids than the female during sexual intercourse. Second, the adolescent female has a higher risk of exposure to sexually transmitted pathogens due to a specific type of cell that lines the cervical canal and other outer surface of the cervix (FHI, 2000a: 7). Third, young girls are often forced to take much older husbands, who are more likely to be infected with HIV/STDs. Fourth, little or no education, unemployment, poverty, family, or husband abuse or violence force many young women to run away from their homes and become sex workers, putting themselves at greater risk of acquiring HIV/STDs because they often have sex with older men who have a long history of sexual activity and, hence, have a higher likelihood of HIV infection. In most cases, young sex workers do not insist on condom use with their clients for fear of loss of income or aggression from the client. Fifth, sexual coercion and other types of violence against women, child marriage, polygamy, abduction, and incest—common occurrences in Ethiopia— put young women at a greater risk for HIV infection than young men. Finally, in urban areas, more and more young people are becoming sexually active at younger ages than in the past. Women who begin sexual activity at an earlier age are more likely to have sexual encounters with high-risk older partners or multiple partners. Many schoolgirls are also frequently forced into sex with older men in exchange for money to support their families or to buy personal needs.

While AIDS has spread throughout the country, only about 84,000 cases were reported to the MOH in the year 2000. Many factors contribute to the low reporting, according to the MOH. First, many people never seek medical care for AIDS because of the stigma attached to the disease. Second, doctors fail to inform AIDS patients of their diagnosis for fear of devastating the patient: some doctors lack the ability to provide counseling and medical support. Third, many people die of other diseases (tuberculosis and fungal infections, for example) before they are diagnosed with AIDS. Fourth, laboratory facilities for diagnosing HIV/STDs are often unavailable at health facilities located outside the major urban areas (MOH, 2000c: 6). Even where testing facilities are available, many people are unwilling to be tested because they know that they are not going to get treatment and that all they will learn is that they are going to die.

The culture of concealment has contributed to the spread of HIV/AIDS in Ethiopia. Many people who believe they are HIV-positive nonetheless avoid getting tested. Instead, they continue perilous practices and spread the disease. In addition, many who know that their partners are HIV positive hesitate to present themselves for HIV testing. Few people speak openly about the disease: the climate of fear and intolerance surrounding the disease discourages frank discussion (Shinn, 2001). The HIV/AIDS study of Kidanu and Banteyerga (2003) shows that religion, the media, and the family, friends, and neighbors of AIDS victims have stigmatized the disease. Religion regards the disease as a punishment by God of human wickedness. The media has turned AIDS into a stigma by disseminating terrifying images of the disease. Family members who fear infection and loss of status defame their sick kin. Neighbors gossip. Some health facilities neglect AIDS patients because the disease is terminal and infectious. Schoolchildren with infected parents are the butt of insults. *Kebeles* and *idirs* ostracize people with HIV/AIDS.

AIDS has done irremediable damage to Ethiopian society. In urban areas, AIDS attacks skilled and educated young people—the very people the country needs for development. The disease has killed many women and men in their prime and has had a devastating impact on life expectancy. Originally, population experts had projected the country's life expectancy to reach 64 years by the year 2010, up from 52 in 2001.With the spread of HIV/AIDS, experts now anticipate that life expectancy will slide back well below 50 years. Experts predict that the child mortality rate will also increase as more infants are born with HIV.

AIDS poses a difficult challenge to the healthcare system. Because AIDS is a pervasive, expensive disease, it demands a disproportionate amount of already scarce healthcare resources. Often, Ethiopia diverts foreign development healthcare assistance to AIDS programs. HIV/AIDS consumes more and more of the time and energy of doctors, nurses, and other healthcare workers: in 2000, AIDS patients occupied about 42 percent of all hospital beds in the country (MOH, 2000c: 9). Thus, healthcare professional cannot spend an adequate amount of time with patients suffering other diseases. As Ethiopia increases spending on AIDS, the country loses its ability to combat other illnesses. In addition, the country funds the fight against AIDS at the expense of other social programs, such as healthcare services, education, housing, and infrastructure schemes that play an important role in the improvement of quality of life.

Funding for the use of antiretroviral medication poses another challenge to the healthcare system. In spite of recent reductions of the prices of anti-retroviral drugs, prices are still way beyond the means of patients and their families. If Ethiopia were to import or manufacture the antiretroviral drugs at the recently reduced cost of $600 per patient per year, the country would have to budget $1.8 billion to treat all its 3 million people with AIDS, not including hospital costs and other financial allocations associated with treatment of the disease (Ayehunie, 2001).[25] This amounts to more than twenty-five times the 1999 total health expenditure for the whole country. By 2005, AIDS treatment costs—without including the cost of antiretroviral drugs—are expected to consume more than one-third of the government health spending (UNAIDS, 1999: 49).

Some medical experts argue that drug cocktails, even if affordable, will not help overcome AIDS. Existing antiretroviral drugs may not have an effect on the strains of the AIDS virus that plague Ethiopia. Vaccine trials in the West show hope for the future. However, according to Harvard health expert Doctor Seyou Ayehunie, the strains of the AIDS virus responsible for 90 percent of infections in Africa are considerably more virulent than the strains endemic to the West: 'more than 90 percent of current vaccine designs are based on Western type virus strains whereas vaccine trials against the type C African strain, which is responsible for 75 percent of infections worldwide, represent only 5 percent of the trials' (Ayehunie, 2001).[26] Hence, countries like Ethiopia may not be able to count on new vaccines, however promising they may be. Ghana's top AIDS expert, Dr Fred Sai, says that the best way to fight the scourge is 'by education, preventive health measures, and creating better living standards.'[27] In the mean time, society should ensure that people living with HIV/AIDS have 'access to simple antibiotics [to treat opportunistic diseases], pain killers, clean water, and adequate nutrition to assure them a minimum standard of quality of life' (Save the Children, 2001: 17).

HIV/AIDS presents a growing health problem in Ethiopia. However, the country cannot afford to spend all its limited resources to fight this disease and thus ignore the perpetual problems that affect the overwhelming majority of the Ethiopian people: poverty and its attendant diseases. Malaria, tuberculosis, diarrhea, and childhood diseases combined kill a significantly larger number of people than AIDS.[28] These diseases are preventable, treatable, and potentially curable; they can be fought through hygiene, clean water, sanitation, and simple medicines. But, unfortunately, the country has not been able to allocate sufficient financial and medical resources to fight them. The focus on the AIDS problem overlooks wider issues about poverty, deprivation, and healthcare deficiencies in the country.

Not only does AIDS pose a financial challenge to the country, but it also devastates families and communities. AIDS causes family expenses to increase sharply. Hospitalization costs many times the patient's or the family's income. The orphans of AIDS victims impose a major financial burden on extended families and whole communities.[29] Rather than attend school many children must seek income to support parents with HIV/AIDS. Rather than focus only on the needs of AIDS patients, Ethiopia must complement its medical efforts against the disease with social programs. As Paul S. Zeitz, executive director of the Global AIDS Alliance, puts it: 'An effective strategy to combat AIDS must also go beyond purely medical approaches. It should include better access to clean water and good nutrition and a range of community-based support services intended to meet the needs of affected families and the millions of children left behind by the AIDS epidemic' (Zeitz, 2003).

The federal government's 1998 HIV/AIDS policy promised first to control the spread of HIV/AIDS through education of the youth about HIV/AIDS in school health classes at all levels beginning from the primary level, education of the public about HIV/AIDS and safe sex via the mass media. The policy also promised to support those infected and affected by the disease, to provide voluntary HIV testing and counseling, and to generate HIV sentinel surveillance data to monitor trends in infection and to direct efforts to slow the spread of the disease. The policy intended to mobilize governmental, NGOs, bilateral and multilateral agencies, the private sector, and local community organizations to provide financial assistance and counseling to HIV/AIDS patients and their families and to AIDS orphans. Finally, the policy promised to protects the rights of people living with HIV/AIDS to access employment, education, housing, and public and private entertainment facilities (FDRE, 1998: 20-34). Five years after the introduction of the policy, however, many of its intended measures have yet to occur.[30]

The government's preventive efforts have thus far focused on increasing public awareness and promoting responsible behavior using broadcast media, workshops, and school education. The government has been able to solicit financial and material support from international organizations, donors, and non-governmental organizations to prevent the spread of the disease by working through schools and health facilities, the workplace, and the media. Unfortunately, however, the government has not been able to expand access to health services including HIV and STD prevention, testing and counseling (which can help reduce HIV risk behavior), and care and support services to people living with HIV/AIDS.[31] By and large, the government has yet to demonstrate the political will necessary to curb the spread of HIV/AIDS (Shinn, 2001; UNOCHA, 2002). TV and radio advertisements and warnings posted on billboards cannot alone contain the spread of the disease because such media outlets do not reach the vast majority of the population that live in rural and small urban areas. Inexpensive interventions could significantly reduce new infection: for example, widespread sex education at schools and beyond, making condoms more available[32] and reasonably priced, treating other sexually transmitted infections (which magnify the risk of ensuing infection with HIV), HIV testing and counseling, taking measures to reduce mother-to-child transmission, and ensuring a safe blood supply (UNAIDS, 2000; World Bank, 1999).

There is a wide consensus among local and international groups involved in the fight against HIV/AIDS that the government can and must do a lot more than it has done thus far. HIV transmission rates have slowed down in countries where governments have put in place effective health and prevention programs. One of the best examples in Africa is Uganda, where HIV infection rates have declined significantly because of combined preventive efforts with access to treatment and care and an aggressive public campaign that promotes safe-sex practices. Political commitment at the highest level of government coupled with openness about the disease have also contributed to Uganda's success. As a result,

HIV prevalence in Uganda fell from 30 percent in 1992 to 5 percent in 2002 (Ocama and Lee, 2003).[33]

While government supports to people living with HIV/AIDS remains minuscule, several NGOs, multilateral and bilateral donor agencies provide funds for HIV activities in large urban areas. A number of local NGOs have begun to mobilize against HIV/AIDS. The Organization for Social Services for AIDS (OSSA), for instance, provides home-based care, counseling and support to people with HIV/AIDS and their families in Addis Ababa, Awasa, Bahir Dar, Dessie, Harar, Mekele, Adama, and Nekemte. OSSA provides support to over 500 anti-AIDS clubs around the country that give information about the disease (OSSA, 1999). Dawn of Hope, a charity organization established by HIV-positive people, has lobbied government leaders for the import or manufacturing of cheap generic life-prolonging antiretroviral drugs. The generic drugs, copies of much more expensive drugs manufactured and patented by the world's leading pharmaceutical companies, have cut AIDS deaths considerably in the West. Dawn of Hope also gives AIDS a 'human face': its members share their personal experiences with the public. The organization fights discrimination against its members in housing, employment, and access to public services, bringing AIDS into the open and encouraging a compassionate response from the government and the society at large. Dawn of Hope also seeks financial support from NGOs and international funding agencies to alleviate the impact of AIDS on its members and their families. The Mekdim Ethiopia HIV Positive Persons and AIDS Orphans National Association also provide wide ranging HIV/AIDS services. The Hiwot HIV/AIDS Prevention, Care, and Support Organization, established in 1999 by Sister Tibebe Maco, educates the public about the modes of HIV transmission and preventive methods and gives care and support to people living with the virus, bedridden AIDS patients, and children orphaned by AIDS in *Wereda* 23, Addis Ababa.[34] There are also a few religious- and community-based organizations that are providing HIV/AIDS prevention, care, and support in various city neighborhoods. All these activities are largely concentrated in and around major cities and, as a result, have little or no impact on rural populations. Moreover, financial constraints prevent commendable local organizations from reaching more than a minuscule proportion of AIDS patients and orphans in urban areas.

The epidemic thus outpaces all efforts to halt it. At present AIDS spreads virtually unimpeded. Only an alteration in behavior on the part of the citizens of Ethiopia can reduce the incidence of HIV/AIDS. Education can encourage people to alter their behavior, making them realize the importance of their own health and the health of others. Those who already suffer from AIDS have a profound influence on the public and can thus play an important role in education about the disease.

Sex education plays a vital role in the modification of youth sexual behavior. Contrary to the beliefs of many, sex education does not lead to earlier or increased sexual activity. Rather, adolescent health experts argue, young people 'are interested in sex because of biological reasons, hormones' (FHI, 2000b: 10) In addition to this, 'suggestions about sex in music, radio, advertisements, films, and television reinforce' sex interest (FHI, 2000b: 10). At any rate, adolescent sexual activity is already high. A 1991 study by the Ministry of Labor and Social Affairs found that 48.5 percent of male and 42.5 percent of female adolescents in urban Ethiopia had their first sexual intercourse at the age of 15-17 and 33.4 percent of male and 25.8 percent of female adolescents in rural Ethiopia, at the ages of 18-19 (MOLSA, 1991).[35] Therefore, sex education is necessary. The experiences of many countries reveal that sex education becomes effective when it is introduced early in primary school and continued through adolescence by teachers who are well trained, and when it involves parents, administrators and religious leaders, and community. Successful curriculum includes 'information on human development, reproductive anatomy, relationships, personal skills, sexual behavior, and gender roles' (FHI, 2000b: 10-11). At present the country lacks an all-inclusive HIV/AIDS education curriculum.

Outside the school system, every sector of the society should promote education regarding the prevention, care, and impact of the epidemic. The public health sectors should not be left with the sole responsibility for dealing with the disease. Trade unions, private and public establishments, religious institutions, traditional self-help associations and other civil organizations must be encouraged and supported to participate in the ongoing war against the disease. Of course, it should be the responsibility of the public health institutions at all levels to perform HIV testing, provide HIV/STD prevention services and education, and treat opportunistic infections. At present, the availability of HIV tests and counseling outside the few large cities is largely non-existent. The lack of testing means that the disease will continue to spread as those infected knowingly or unknowingly pass it to their sexual partners.

*Leading causes of morbidity and mortality* In spite of modest public efforts to improve the healthcare system, the pattern of morbidity and mortality has changed little. The country has not been able to control or prevent the communicable diseases and the bacterial and parasitic diseases associated with poverty and poor physical environments, such as diarrheal diseases, respiratory infections, and malaria.

Tables 6.5a, b, and c summarize the leading causes of outpatient visits, hospital admissions, and death, respectively, for the whole country. Malaria, pneumonia, tuberculosis of respiratory system, acute upper respiratory infection, and helminthiasis appear to cause much misery in Ethiopia.

Malaria, a water-related insect vector-borne disease, interpenetrates three-fourths of Ethiopia; two-thirds of the population risk infection.[36] The disease occurs with the greatest frequency in the warm lowlands, below 1,500 meters altitude with a mean annual temperature of 20-30 degrees Celsius and rainfall total of 100 to 900 millimeters. The lowlands count for half of Ethiopian land and host many of the country's communicable diseases. In the temperate areas, between 1,500 and 2,500 meters altitude with a mean annual temperature of 20 degrees Celsius and rainfall of 400 to 2,400 millimeters, malaria occurs less often but with a greater morbidity: outbreaks occur mostly around 2,000 meters altitude, producing a higher mortality rate than in the lowlands because the inhabitants do not build up an immunity from continual communal exposure. Malaria does not affect the cold zone, above 2,500 meters altitude with a mean annual temperature of 15 degrees Celsius and rainfall total of 1,000 to 1,600 millimeters). Ethiopia hosts four vectors of malaria. *Anopheles arabiensis*, the most common vector, breeds in sunlit puddles during the rains in most regions of Ethiopia. *Anopheles pharoensis*, the second most common vector, breeds along the rivers of Baro and Awash and the Rift Valley lake region and along the Tana Lake. *Anopheles fenestus*, the third most common vector, also breeds along the shores of lakes and large rivers. *Anopheles nile*, the least common vector, breeds in the southwestern, western, and northern low altitude parts of the country (MOH 2001: 5).

Malaria is the leading cause of hospital inpatients and outpatients in Ethiopia. It was the leading cause of death in six of the nine regional states (Afar, Amhara, Benishangul-Gumuz, Gambela, SNNPR, and Tigray) in 1999. The parasite is transmitted between humans by female mosquitoes. The *Plasmodium* completes its several asexual lifecycle stages in the human liver and in red blood cells and then undergoes sexual development and sporogony in the mosquito (WHO/WMO/UNEP, 1996: 78). Of the four types of *Plasmodium* species (*Plasmodium falciparum, vivax, ovale,* and *malariae*), falciparum—the most virulent species clinically—accounts for the majority of the infections and causes nearly all malaria-related deaths. Of the patients diagnosed with malaria between 1995 and 1999, about 62 percent contracted *Plasmodium falciparum* while almost all the remaining cases were *Plasmodium vivax*. Tigray, Afar, Somali, Benishangul-Gumuz, and Gambela were the worst hit with over three-quarters of the malaria cases attributed to *Plasmodium falciparum* (MOH, 2000a: 30). *Plasmodium malariae*, most frequently reported from the Arba Minch areas, causes less than one percent of all cases. The incidence of *Plasmodium ovale* is negligible. The emergence of anti-drug resistant strains has exacerbated the malaria problem. The disease is back in areas where it has previously been controlled and has become more severe than in the past. The disease has made a drastic resurgence due largely to ineffective vector and disease surveillance, inadequate public health infrastructure, and population growth-induced migration of people into endemic areas.[37]

In 1998, the WHO launched the 'Roll Back Malaria' initiative with the cooperation and partnership of United Nations and development agencies, the World Bank and the International Monetary Fund, governments, the private sector, researchers, and NGOs to combat malaria mortality and morbidity worldwide. The new initiative involves 'focusing on early detection and prompt treatment of cases of malaria, the detection and control of epidemics, vector control using bed nets treated with insecticide, and the prevention and treatment of malaria in pregnancy.'[38] By the end of the year 2001, three years after the Roll Back Malaria initiative was launched, very little of the plans have been implemented at ground level.[39]

**Table 6.5a  Top 10 Leading Causes of Outpatient Visits, 2001**

| Rank | Diagnosis | Cases | % |
|---|---|---|---|
| 1 | All types of Malaria | 328,760 | 10.4 |
| 2 | Helminthiasis | 213,195 | 6.7 |
| 3 | Acute upper respiratory infection | 205,129 | 6.5 |
| 4 | Bronchopneumonia | 173,123 | 5.5 |
| 5 | Infections of Skin & Subcutaneous tissue | 145,680 | 4.6 |
| 6 | Gastric & Duodenites | 137,942 | 4.4 |
| 7 | Dysentery | 111,938 | 3.5 |
| 8 | Tuberculosis of respiratory system | 70,526 | 2.2 |
| 9 | Sexually transmitted infection | 68,733 | 2.2 |
| 10 | Bronchitis, Chronic and unqualified | 58,594 | 1.9 |
|  | Total of the above cases | 1,513,620 | 47.8 |
|  | Total of all cases | 3,167,514 | 100.0 |

**Table 6.5b  Top 10 Leading Causes of Hospital Admissions, 2001**

| Rank | Diagnosis | Cases | % |
|---|---|---|---|
| 1 | All types of Malaria | 16,782 | 14.8 |
| 2 | Pneumonia | 10,090 | 8.9 |
| 3 | Tuberculosis of Respiratory System | 8,881 | 7.8 |
| 4 | Accidents | 6,967 | 6.2 |
| 5 | Abortion | 4,449 | 3.9 |
| 6 | Pregnancy, childbirth & the Puerperium | 4,326 | 3.8 |
| 7 | Cataract | 2,735 | 2.4 |
| 8 | Bacillary Dysentery | 1,848 | 1.6 |
| 9 | Gastroenteritis & Colitis | 1,707 | 1.5 |
| 10 | Meningitis | 1,015 | 0.9 |
|  | Total of the above cases | 58,800 | 51.9 |
|  | Total of all cases | 113,365 | 100.0 |

**Table 6.5c  Top 10 Leading Causes of Deaths, 2001**

| Rank | Diagnosis | Cases | % |
|------|-----------|-------|---|
| 1 | Tuberculosis of Respiratory System | 1,005 | 10.1 |
| 2 | Pneumonia | 734 | 7.3 |
| 3 | All types of malaria | 462 | 4.6 |
| 4 | Bacillary dysentery | 224 | 2.2 |
| 5 | Accidents | 156 | 1.6 |
| 6 | Meningitis | 149 | 1.5 |
| 7 | Hypertension | 142 | 1.4 |
| 8 | Gastro-enteritis & Colitis | 109 | 1.1 |
| 9 | AIDS | 83 | 0.8 |
| 10 | Leishmaniasis | 48 | 0.5 |
| | Total of the above cases | 3,113 | 31.1 |
| | Total of all cases | 10,006 | 100.0 |

Source: Ministry of Health (2001), *Health and Health Related Indicators*, Addis Ababa: MOH, p. 33.

Ethiopia is one of eight African countries where tuberculosis has made a rapid come back.[40] It is a disease of the lungs and other organs caused by *Mycobacterium tuberculosis*. Nine out of ten cases of the disease occur in developing countries (Kusner, 1993: 280). About one-half of Ethiopian adults carry a dormant tuberculosis infection, which is suppressed by the immune system (MOH, 2000c: 31). In 2001, tuberculosis of the respiratory system killed more hospital and health center patients than any other diseases and it was also the third leading cause of admission to hospitals and health centers (MOH, 2001: 33). Tuberculosis has become the leading cause of death among people with AIDS.[41] Pneumonia was the second cause of hospital morbidity and mortality in 2001. The diarrheal diseases continue to be a major public health problem in the country, largely because of the lack of adequate sanitation. They are one of the most common causes of morbidity of infants and young children. Acute childhood diarrhea (ACD) is the leading cause of death in children (3 years old and younger) and accounts for over 300,000 deaths in the country annually (Ethiopian Health and Nutrition Research Institute, 2001: 3).[42] Children in Ethiopia come down with diarrhea from 2 to 6 times each year; annual diarrhea episodes among children under five years of age are estimated at 50 million (UNICEF, 1997: 13). The *Shingella* bacteria, which causes dysentery or blood diarrhea is responsible for most diarrhea deaths (WHO, 2001b: 6). The overall deprived social and economic condition of the population, poor living conditions, lack of good nutrition, safe drinking water and sanitation, and little or no access to health services contribute to the high burden of mortality due to tuberculosis, pneumonia, malaria, and diarrheal diseases.

## Traditional Medical Systems

In Ethiopia, traditional healers continue to play a significant role in the treatment of illness, especially in rural areas. The national *Report on the 1998 Welfare Monitoring Survey* indicated that about 5.4 percent of rural and 2.2 percent of urban medical care seekers consulted traditional healers in that year (CSA, 1999: 54, 56). The figures cannot account for all clients of traditional healers, because most people will not openly talk about their use of traditional medicines. Traditional healers elude classification. Some practice medicine as their principal occupation; others practice it as a part-time job. Some perform as specialists working on bone-setting, faith healing, or as practitioners for particular diseases, while others claim to cure a host of illnesses. Some traditional healers may have apprenticed family forerunners; others may practice without any training on the basis of some natural gift or talent. Ethiopian traditional medicines utilize growing plants, holy water, minerals, herbs, roots, grains, spices, oil seeds, butter, blood, fat, cow-dung, urine, prayers, and magic (Vecchiato, 1993: 166; Desta and Meche, 1999: 12).

Past and present Ethiopian governments have had a forbearing attitude toward traditional medicine. Article 518 of the Penal Code of 1957 makes 'unlawful exercise' of public medicine punishable, but the fourth clause of this article allows sufficient room for traditional healers to practice their trade within their communities. The clause reads: *Nothing in this article shall prevent the practice of a system of therapeutics according to indigenous methods by persons recognized by the local community to which they belong, to be duly trained in such practice, provided that no such person shall be authorized to practice except amongst the local community to which he belongs and in such a manner as is neither dangerous nor injurious to the person's health or life.*[43]

Even though the MOH has recognized the importance of traditional medicine since 1975, little has been done to evaluate and incorporate such services. During the Derg regime, traditional birth attendants were trained and incorporated into the healthcare service. A few medicinal plants were also researched for possible use but nothing positive came out of such endeavors (Vecchiato, 1993: 157).

For economic reasons, traditional medical practitioners guard their knowledge: most do not want to share their knowledge with modern medical establishments, because they fear that modern doctors will steal their product and deprive them of their livelihood. Hilary Nwokeabia has conducted studies elsewhere in Africa that show three reasons for the secretive behaviors of traditional medical practitioners: 'the inadequacy of perceived financial return from innovation; lack of public protection of intellectual properties; and the threat that other individuals will illegally appropriate their knowledge' (Nwokeabia, 2002: 40). Her study also found that 75 percent of the traditional medical practitioners would share their knowledge with modern science in exchange for a monetary reward. Ninety percent of those who would share their knowledge in exchange for money believe that the recording of their work for future generations would be an excellent thing. Two percent of traditional medical practitioners would share their work with no promise of a monetary reward. Thirty-six percent of the sample refused to violate the traditional requirement of secrecy regardless of the incentive (Mwokeabia, 2002: 27-28).

Modern medicine may pour scorn on much of the traditional healing practices, but, as Richard Pankhurst put it, 'traditional Ethiopian medical and surgical practices [are] far from arbitrary and while some methods of treatments are useless and even dangerous, others have proved to be useful' (Pankhurst, 1990).[44] Many Ethiopian traditional medical practitioners have succeeded in a variety of medical applications: herbal medicine and bone setting comprise their chief success. Society thus cannot afford to ignore them. Consequently, modern medical practitioners must conduct comprehensive research into their healing practices and activities so that society can benefit from their accumulated knowledge. Financial reward, recognition, and most importantly public protection of their knowledge rights are good incentives for the pursuance of such an action.

## Healthcare Financing

The government plays a major role in directly providing healthcare, in establishing and maintaining facilities, and in employing health professionals. The bulk of healthcare financing—over two-thirds—comes from the general government budget. External assistances (bilateral and/or multilateral, NGOs) contribute anywhere from 15 to 25 percent while user fees put in around 10 percent (Desta and Meche, 1999). In the past few years, the federal and regional governments have been at work to reform the healthcare financing system. The reform effort is specifically aimed at finding ways to redeem part of the costs incurred by the public sector in order to improve the financial problems of the healthcare system. The ongoing reform efforts are looking into three possible ways to recover cost: reducing the rate of waivers or exemptions, increasing user fees, and promoting the idea of health insurance. Public policy-makers in the health sector expect that the additional income derived from undertaking these reforms will go a long way toward solving the severe financial problems of the healthcare system.

*Reducing the rate of waivers or exemptions* At present public health institutions provide free services to most patients. The proportion of patients who receive free medical services is estimated at between 30 and 80 percent. Patients are eligible for a user fee waiver if they bring a letter from their local jurisdiction certifying their neediness. The original idea of exemption was to protect those people who are unable to pay, but many people abuse the system by falsely presenting themselves as destitute, often with the consent of corrupt local public officials. A recent MOH survey shows that a large proportion of patients received user fee waivers on account of their impecuniousness—up to 95 percent at the Dil Chora Hospital in Dire Dawa, 85 percent at Felege Hiwot Hospital in Bahir Dar (Amhara State) and at Mekele Hospital (Tigray State), 70 percent at Adigrat Hospital (Tigray State), 68 percent at healthcare establishments in Hadiya zone (SNNPR State), and more than 60 percent of patients who visited public health centers in Asela Zone (Oromiya State) (MOH, 2000b: 12-26). The government's ongoing healthcare financing reform plans to put in place tougher legal measures to cut down on user fee exemptions. One of the ideas under consideration is to make the local government jurisdictions that issue eligibility certificates to poor patients pick up the bills themselves rather than shift the cost to public healthcare institutions.

*Raising user fees* At present the health delivery system does not depend on user fees. User fees contribute very little to healthcare expenditure. The healthcare financing strategy envisions an increase in charges for health services paid by patients at the point of delivery. The rationale for such a reform is that user fees for health service have remained unchanged for over half a century. A visit to a public health institution now costs a patient 2 Birr or less for registration; it is less than 1 Birr in most cases. Such fees are considered too little to recover the cost of bookkeeping let alone offset the cost of services rendered to the patient. At the present time, the health sector redeems only less than 10 percent of its recurrent healthcare cost via user fees. Income from user fees declined from 19 percent in 1981 to 7.9 percent in 1994 (MOH, 2000b: 1-2). The government's healthcare financing plans intends to reverse this trend, and many regions and cities have already carried out detailed studies of user fee structures and have presented reform proposals to their respective legislators for approval.

In the mean time, as part of the healthcare cost recovery effort, the government has introduced a policy it calls *health facility revenue*. Traditionally all public health facilities remit all revenues (user fees, donations, gifts, and other revenues) to the government coffer. Under the new *health facility revenue* policy, however, public health establishments are allowed to retain all revenues they generate and use these resources toward improving the quantity and quality of health services rendered at their facilities. In other words, in addition to the recurrent budget they receive on an annual basis, public health facilities will have extra revenues that they can put into income-generating endeavors to improve their services. In this regard, the most favored income-earning source is the establishment of special pharmacies that are managed by each public health facility. The special pharmacies are different from those budget pharmacies that are funded annually by the government treasury to render services to poor patients. The sources of funding for the initial capital for the so-called special pharmacies are donors or the government or both and, in some instance, credits from drug companies. Special pharmacies are strictly for-profit enterprises and are allowed to make up to 40 percent profit on drug sales. Special pharmacies have been established in some hospitals in Addis Ababa and Dire Dawa administrative councils, Oromiya, Amhara, SNNPR, Tigray, and Harari regional states and are generally credited for improving drug supplies, providing around-the-clock services, and for using some of the profits to augment other health services in their establishments (MOH, 2000b: 11-26).

Many health experts, however, criticize the attempts to increase user fee charges on several grounds. First, the vast majority of the population is too poor to afford any additional healthcare cost. User fees also prevent the poor from availing themselves of existing healthcare services. Second, the quality of the current healthcare service is so deprived that additional user fees are not warranted. Third, an increase in user fees will push even more people into seeking waivers on account of hardship. Fourth, since taxpayers finance the public healthcare system, they should be entitled to access free health services; hence user fees are totally unnecessary.

Some evidence supports the arguments of user fee critiques. Many studies in Africa and elsewhere in the developing countries in the 1990s found numerous cases of decline in access to healthcare services after user fees had been implemented. For example, after examining the relationship between user fees and access to maternity health services, Murray (1996) found that the introduction of fees led to a marked drop in hospital admissions and deliveries and an increase in the mortality rate of mothers and their babies born outside the hospital.[45]

*Health insurance*   This component of the government's healthcare financing strategy remains at a conceptual level for two main reasons. First, the private health insurance market in the country more or less does not exist. Second, the vast majority of the population has such a low income that it cannot afford any health insurance premium. Only fewer than 10 percent of the labor force in the country is engaged in formal sector employments and, hence, the vast majority of the working population cannot purchase health insurance through their employer. The government's health experts, however, foresee the possibility of promoting collective community health insurance schemes in the near future, especially in urban communities.

## Conclusion

The healthcare system in Ethiopia is woefully inadequate to treat environment-related diseases or to prevent them from occurring. The current level of recurrent and capital expenditures on health cannot ameliorate the growing health problems of the country. In 2000/2001, the government spent only US $1.17 or 11.5 Birr per person on healthcare, 90 percent below the level for the minimum basic health service package recommended by the World Bank (US $12.50). Increases in user charges and reductions in the rate of exemptions hardly boost healthcare expenditures. In fact, administrative costs to run the system will consume most or all of the additional revenue collected through such means. At any rate, public expenditure on health could be significantly augmented without user fee increases if available resources were judiciously utilized. The current government budgetary appropriations are terribly misguided. Expenditures on government bureaucracies, national defense, public order and security, and general services soak up close to three-fifths of the public budget while social and economic services receive no more than one-fifth. Close to two-fifths of the government health expenditures are geared towards covering the administrative cost of the MOH itself. Such imprudent priorities must be reversed.

## Notes

[1] Cited in Annalee Yassi, Tord Kjellstrom, Theo de Kok, and Tee L. Guidotti (2001), *Basic Environmental Health*, Oxford: Oxford University Press, p. 5).

[2] BBC News, 'Hopes Rise for Malaria Vaccine,' 2 December 2002. Available at: http://news.bbc.co.uk/2/hi/health/2529347.stm.

[3] Health service coverage is defined as the presence of a health facility serving a population living within a 10-kilometer radius (Markos Ezra and Seyoum Gebre Silassie (1998), *Handbook on Population and Family Life Education for Secondary School Teachers in Ethiopia*, Addis Ababa: Institute for Curriculum and Development Research, p.31.

[4] Unless otherwise indicated, statistics used in this section are obtained from the Ministry of Health (2001), *Health and Health Related Indicators 2000-2001*, Addis Ababa: MOH.

[5] The government's health policy incorporates the promotion of private sector participation in the provision of healthcare services by providing tax holiday, exemptions from import duty for health equipment, and access to public land at low price (Ministry of Economic Development and Cooperation (MEDaC) (1999), *Survey of the Ethiopian Economy*, Addis Ababa: MEDaC, p. 30).

[6] Walta Information Center, 'Ethiopia Ranks First in the Immigration of Medical Professionals.' Available at: http://www.waltainfo.com/EnNews/2002/Nov/08Nov02/Nov08e2.htm.

[7] United Nations Integrated Regional Information Network, "High Death Rate from Illegal Abortion,' 28 October 2002. About 1,209 women die for every 100,000 women who have abortions in Ethiopia. Available at: http://allafrica.com/stories/200210280407.html

[8] According to a 2003 global study by Save the Children, 'Ethiopia is one of the worst countries in the world to be a mother.' Ethiopia ranks third from the bottom of 117

countries, 'only Niger and Burkina Faso ranking worse.' The ranking is based on six factors, including: 'risk of maternal mortality, women using modern contraception, births attended by trained personnel, pregnant women with anaemia, adult female literacy rate, and participation in national government' (United Nations Office for the Coordination of Humanitarian Affairs (2003), 'Ethiopia: One of world's worst countries for mothers,' 7 May. Available at: http://irinnews.org/report.asp?ReportID=33934).

[9] Cited in WHO (2001a), *Maternal Mortality*, WHO: Division of Reproductive Health (Technical Support): Geneva, p.3.

[10] According to the UNICEF, measles alone kill in excess of 70,000 children every year contributing to about half of the mortality of children (United Nations Office for the Coordination of Humanitarian Affairs, 'Interview with Dr. Mahendra Sheth of UNICEF,' 30 April 2003. Available at:
http://www.irinnews.org/report.asp?ReportID=33784&SelectRegion=Horn_of_Africa&Sel ectCountry=ETHIOPIA).

[11] For almost two years since January 2001, no cases of poliovirus had occurred in Ethiopia and if this trend continues for one more year, the country will be certified polio-free by the World Health Organization (United Nations Office for the Coordination of Humanitarian Affairs, 'Ethiopia: Largest Ever Polio Vaccination Campaign Underway' 25 October 2002. Available at: http://irinnews.org/report.asp?ReportID=30601).

[12] CSA defines chronic malnutrition or stunting as 'nutritional deficient state related to frequent episode of acute malnutrition or a long-term food deficiency, often combined with persistent or recurrent ill health in first few years of life. A child who is less than 90 percent of the reference median of height-for-age, or two standard deviations below it, is classified as chronically malnourished or stunted.' (CSA (1999), *Report on the 1998 Welfare Monitoring Survey*, Statistical Bulletin No. 224, Addis Ababa: CSA, p. 9).

[13] *Marasmus* is a nutritional deficiency disease that results from a diet low in calories and protein. *Kwashiorkor* is a nutritional deficiency disease that occurs in infants and very young children when they are weaned from mother's milk to a starchy diet that is relatively high in calories but low in proteins (G. Tyler Miller (2002), *Living in the Environment*, Belmont: Wadsworth Group, Glossary, G7-8).

[14] Support for FGM is heavily influenced by level of education and place of residence. Women with secondary and tertiary levels of education are less likely to support it (18.6 percent) than women with no education (67 percent). Urban women are less likely to support the practice (31 percent) than rural women (66.1 percent) (CSA (2001a), *Demographic and Health Survey*, Addis Ababa: CSA, p. 33).

[15] The following is a brief summary of the WHO's excellent study on the consequences and the long-term complications of the different types of FGM (WHO, 1998: 14-30):

*Death* can result from severe bleeding (hemorrhagic shock), from the pain and trauma (neurogenic shock) or from severe and overwhelming infection (septicemia). *Hemorrhage*: severe bleeding is the most common immediate complication. *Shock*: the child may enter into a state of shock from the pain, experience psychological trauma and exhaustion from screaming. *Injury* to neighboring tissue such as to the urethra, the vagina, the perineum or the rectum can lead to the formation of fistulae through which urine or feces will leak continuously. *Urinary retention*: pain, swelling and inflammation around the wound and subsequent infection can lead to urine retention, which may last for hours or days, but is usually reversible... by removing the stitches. *Infection* is very common and can be caused by an unsterilized instrument... and the area becomes soaked in urine and contaminated by feces...unsterilized tools and fecal matter can cause infection with tetanus spores or bacteria that will cause gangrene. *Severe pain*: the majority of procedures are performed without an anesthetic. When local anesthesia is used, pain in the highly sensitive area of the clitoris returns within 2-3 weeks of operation. Applying the local anesthesia is itself extremely painful because the area of the clitoris and labia minora has a dense concentration of nerves and is highly sensitive.

Long-term complications of types I and II may include: *Failure to heal*: scar over the clitoris may split open during childbirth. This may lead to profuse bleeding from the clitoral artery. *Abscess formation*: In case where the infection is buried under the wound edges or an embedded stitch to be absorbed, an abscess can form which will usually require surgical incision and repeated dressing over a period of time. *Dermoid cyst*: This is the most common long-term complication of all types of FGM. It results from the embedding of skin tissue in the scar. The gland that normally lubricates the skin will continue to secrete under the scar and form a cyst or sac full of cheesy materials.... If cysts become very large or infected, surgical removal may be unavoidable. *Keloids*: There is a genetic susceptibility to keloids (excessive growth of scar tissue) in many of the ethnic groups that practice FGM. Vulval keloids are disfiguring and psychologically distressing. *Urinary tract infection*: presence of pus and infection near the short urethra can cause recurrent ascending urinary tract infections. *Scar neuroma*: Trapping of the clitoral nerve in stitch or in the scar tissue of the healed wound can result in a neuroma (tumor consisting of neural tissue). This can make touching the vulva during sexual intercourse or washing very painful. *Painful sexual intercourse*: Sexual intercourse can become painful and psychologically distressing as a result of one or several of the complications mentioned above.

Long-term complications of type III FGM: As cutting is more extensive in type III FGM, the long-term complications include those mentioned for types I and II and: *Reproductive tract infections*: Ascending infections from the vulva due to retained discharge and blood can lead to pelvic inflammation disease. *Dysmenurrhoea* (painful menstruation). *Chronic urinary tract obstruction*: A tight infibulation or urethral stricture resulting from accidental injury can cause obstruction of urethral flow, repeated infections and bladder stones. *Urinary incontinence*: Dribbling of urine is common in infibulated woman; the bladder is not completely emptied and chronic infection under the hood of scar tissue makes control of urine difficult. *Stenosis of the artificial opening to the vagina*: With infibulation, the artificial opening of the vagina can be so small that it closes almost completely overtime. This may cause incomplete voiding of urine or haematocolpos (retained menstrual blood) and make sexual intercourse impossible. *Complications in labor and delivery*: During childbirth, the infibulated woman must be defibulated to allow the fetal head to emerge from the vagina. This increases the risk of bleeding and wound infection. If an experienced attendant is not available to perform defibulation (anterior epistiotomy), labor may become obstructed. Prolonged obstructed labor can cause moderate-to-severe complications for the mother and child.

[16] The Addis Ababa Fistula Hospitals estimates that for every woman who dies of pregnancy complications, 10 will develop fistulas (The International Planned Parenthood Federation, 'Ethiopia: Specialized Surgery Offers Cure for Fistula.' IPPF-News, 28 October 2002). The maternal mortality rate in Ethiopia is among the highest in the world and is estimated at between 560 and 850 per 100,000 live births.

[17] According to the 2000 *Demographic and Health Survey* (CSA, 2001a: 32), a large majority (84.5 percent) of women believe that a husband is justified to beating his wife for at least one of the following reasons: if she burns the food (64.5 percent), argues with him (61.3 percent), goes out without telling him (56.2 percent), neglects the children (64.5 percent), or refuses sexual relations (50.9 percent).

[18] A 2000 USAID-funded report on AIDS orphans estimated that the number of children orphaned by AIDS in Ethiopia rose from 2.3 percent in 1990 to 25.8 percent in 2001. The report projects that this percentage will rise to 35.4 percent in 2005 and 43 percent in 2010 (UNAIDS and UNICEF (2002), *Children on the Brink 2002: A Joint Report on Orphan Estimated and Program Strategies*, Washington, D.C.: USAID).

[19] BBC World Service, 'AIDS Demo By Ethiopian Orphans.' Available at: http://news.bbc.co.uk/2/hi/africa/1474917.stm. Accurate data on AIDS cases—that is, the number of AIDS case by age, gender, and transmission category—are non-existent. Existing HIV data are mostly drawn from official surveillance figures gathered from specific groups, including women who attend antenatal clinics, people who visit health facilities for sexually transmitted diseases (STDs), people with tuberculosis, blood donors, sample blood tests from students, commercial sex workers and soldiers, and prevalence reports by academic health professional found in health journals. The 2000 national *Demographic and Health Survey*, which provides important information on level of awareness about AIDS, modes of transmission, levels of condom use, and attitudes towards the disease, was designed to track HIV infections in the general population. While data obtained from these groups and sources are very important to generate public response to HIV, target prevention activities, and raise awareness about the epidemic, they do not represent the general population. Hence, the HIV/AIDS data cited in this section are estimates and subject to bias.

[20] Yohannes Ruphael, 'The Challenges Faced by our Women.' Available at: http://www.addistribune.com/Archives/2004/03/19-03-04/Challenges.htm.

[21] The AIDS virus is also spreading in rural areas due to increased rural-urban and urban-rural mobility. The spread of the virus to small towns and rural areas may have been facilitated by refugees and disbanded soldiers who return to their homes after the end of the civil war in 1991.

[22] Many soldiers contracted HIV during the civil war in the 1980s by having contact with multiple sex partners. When the civil war ended in 1991, thousands of infected soldiers returned home spreading HIV/AIDS in their towns and rural communities. Ethiopia is one

of five countries (the others being Nigeria, China, India, and Russia) with fastest rising rates of infections from the AIDS virus (Lawrence K. Altman (2002), 'AIDS in 5 Nations Called Security Threat,' *The New York Times*, 1 October).

[23] These reported rates are not for all commercial sex workers in these urban areas, but only for those sex workers who volunteered to be tested for HIV.

[24] According to the 2000 *Demographic and Health Survey* (CSA, 2001a: 75), polygyny is widely practiced in the Gambela Regional State (29 percent), Afar Regional State (24 percent), the SNNPR Regional State (22 percent), and Oromiya and Somali (18 percent each). Only 2 percent of married women in the Amhara Regional State and Addis Ababa were in a polygynous relationship.

[25] In April 2003 Ethiopia became one of the first African countries, alongside South Africa, scheduled to receive technological support to manufacture its own anti-retroviral (ARVs) drugs to tackle the devastating HIV/AIDS crisis. But it was unclear at that point who was going to 'provide the technology to manufacture generic medication in the country or how those drugs' are then going to be distributed and at what price (United Nations for the Coordination of Humanitarian Affairs, 'Ethiopia: Anti-HIV/AIDS drugs to be produced locally.' Available at: http://www.irinnews.org/report.asp?ReportID=33491.

[26] Globally, only 10 percent of medical research is devoted to the diseases that cause 90 percent of the world's health burden. Most medical research is focused specifically on diseases affecting the most developed world (Lawrence K. Altman, 'Gates Foundation Donates $200 Million.' Available at:
http://www.nytimes.com/2003/01/27/national/27GRAN.html). In the last 15 years or so, international drug companies have developed 1,700 medicines approved for clinical use. However, only 11 were aimed at tropical diseases (Adam Lusekelo, 'Africa's aid plan seeks healthy growth.' Available at:
http://news.bbc.co.uk/2/hi/business/2819885.stm (*BBC News*, 6 March, 2003).

[27] Quoted in "It Takes a Village" *The New York Times*, 27 April 2001, Op-Ed, by Thomas L. Friedman.

[28] In fact, people infected with the HIV do not die of AIDS. They die of tuberculosis, pneumonia, internal hemorrhaging or other diseases as the natural defense system of the body is weakened by AIDS (Vandana Shiva (1994: 21), 'Women, Ecology and Health: Rebuilding Connections,' in *Close to Home: Women Reconnect Ecology, Health and Development Worldwide*, Vandana Shiva (ed.), Philadelphia: New Society Publishers).

[29] The Ministry of Health projects that the number of orphans living in Ethiopia will increase to 1.8 million in 2007 and to 2.5 million in 2014 (Ministry of Health (2002), *AIDS in Ethiopia*, Addis Ababa: Ministry of Health). A 2000 study of AIDS orphans in Bahir Dar, capital of the Amhara Regional State, by Dr. Marta Segu and Dr. Sergut Wolde-Yohannes of the Boston University School of Public Health revealed that AIDS orphans lacked adequate shelters, food and clothing, employment opportunities, health services, and educational opportunities and suffered stigmatization, rejection, and physical and mental abuses (Marta Segu and Sergut Wolde-Yohannes (2000), 'A Mounting Crisis:

Children Orphaned by HIV/AIDS in Semi-Urban Ethiopia,' in *Orphan Alert: International Perspectives on Children Left Behind by HIV/AIDS*, Lutry, Switzerland: Association Francois-Xavier Bagnoud, July).

[30] This is in spite of the government's success in the 1990s in acquiring substantial amount of funding from bilateral and multilateral organizations for AIDS-related projects. Donor support for HIV/AIDS in Ethiopia in 2002 included: Christian Relief and Development Agency (US $233,400 for 2001-2003 for HIV/AIDS intervention, NGO capacity building, and advocacy); World Health Organization (US$1.75 million for 2002-2003 for technical support to health sector HIV/AIDS intervention); UNICEF (US $3.4 million for 2001-2002 for support to health sector HIV/AIDS, PMTCT, and youth prevention); UNAIDS (US $486,000 for 2002-2003 for support of HAPCO, Support Ministry of Health in VCT); United Nations Development Program (UNDP) (US $3 million for 2002-2006 for leadership, advocacy and communication, mainstreaming in planning, human rights, socioeconomic research); World Bank (US $13.36 million for 2001-2002 for funding for the Ethiopian Multi-sectoral AIDS Program); USAID (US $8.2 million for FY2001 for prevention, care and support); Ireland Aid (US $34,000 for 2002 for support to Ministry of Health VCT); Netherlands (US $1 million for 2002-2003 for Ethio-Netherlands AIDS Research Project); Norway (US $1.3 million for support channeled through UNICEF and HAPCO); German Agency for Technical and Cooperation (GTZ) (US $120,000 for 2002 for support to Amhara, Tigray, and Oromiya Regions). The World Bank, the United Nations Development Program (UNDP), the Joint United Nations Program on HIV/AIDS, United Nations Children's Fund, United Nations Educational, Scientific, and Cultural Organization (UNESCO), and the United Kingdom are major donors for HIV/AIDS interventions in Ethiopia (Global Fund Coordinating Committee of the Federal Democratic Republic of Ethiopia, *The Global Fund Proposal to Reduce Malaria, HIV/AIDS and TB in Ethiopia*, Geneva, June 2002). Available at:
 http://www.globalfundatm.org/fundingproposals/ethiopiahivuk.doc; Cited in Lisa Garbus, *HIV/AIDS in Ethiopia*, AIDS Policy Research Center, University of California San Francisco, April 2003, pp. 81-82).

[31] In May 2003, the government announced a plan to extend countrywide services to help prevent mother-to-child transmission of the disease, which at that point was only available in Addis Ababa (United Nations for the Coordination of Humanitarian Affairs (2003), 'Ethiopia: UNICEF, government Tackling HIV/AIDS Transmission to Children,' 3 May).

[32] By and large, condom use in Ethiopia is low. According to the 2000 *Demographic and Health Survey* (CSA, 2001a: 175), knowledge of condoms is not widespread, especially among women. Only 35 percent of all women know about condoms while the comparable figure for men was 68 percent. Only 12 percent of women knew of a source for condoms.

[33] Uganda's investment of debt relief money for its national HIV/AIDS plan also played a key role in the government's success in reducing HIV infection rates.

[34] Sister Tibebe won the 2001 Hunger Project Africa Prize (US$50,000) for her humanitarian work in fighting to stop the spread of AIDS.

[35] Cited in Markos Ezra and Seyoum Gebre Silassie (1998), *Handbook on Population and Family Life Education for Secondary School Teachers in Ethiopia*, Addis Ababa: Institute for Curriculum and Development Research, p.55.

[36] *Walta Information Center*, Addis Ababa, 1 May 2001.

[37] There is currently no vaccine against malaria, but scientists are said to be close to making a breakthrough in developing a vaccine that can successfully protect people against the disease. (BBC News, 'Hope Rise for Malaria Vaccine,' 2 December, 2002. Available at: http://news.bbc.co.uk/2/hi/health/2529347.stm)

[38] Gavin Yamey (2001), 'Global campaign to eradicate malaria,' BMJ: 322, 1191-1192 (19 May). Available at: http://www.bmj.com/cgi/continent/full/322/7296/1191.

[39] In April 2000, 43 African nations signed the Abuja Declaration, pledging to waive or reduce tariffs and taxes on mosquito nets and anti-malarial drugs and endorsed the Rollback Malaria Campaign's goal to halve the number of cases by 2010. However, 26 African nations—including Ethiopia—had yet to live up to their pledge by the end of 2002. Tariffs and taxes increase the cost of mosquito nets by at least 40 percent (Blaise Salmon 'Malaria Bednet Taxes Must Go.' Available at: http://www.addistribune.com/Archives/2003/03/07-03-03/Malaria.htm).

[40] Ethiopian Radio, March 23, 2002. With 397 incidences per 100,000 people, Ethiopia has one of the highest rates of tuberculosis occurrence (World Bank, 2003: 22).

[41] Because tuberculosis is a contagious disease, it has the potential to infect people that are not HIV positive. This will, in turn, increases the number of people requiring medical attention, thus putting a strain on the country's healthcare infrastructure.

[42] Prevention of ACD depends upon the availability of safe drinking water and the improvement of environmental sanitation. However, most Ethiopian children do not have access to clean water, nor do they live in sanitary conditions.

[43] Cited in W.F.L. Buschkens and L.J. Slikkervee (1982), *Health Care in East Africa: Illness Behavior of the Eastern Oromo in Hararghe (Ethiopia)*, Van Gorcum Assen, The Netherlands, p. 37.

[44] Cited in Asfaw Desta and Hailu Meche (1999), "Review of the Development of Health Services in Ethiopia," in *Proceedings of the 10th Annual Scientific Conference of the Ethiopian Public Health Association, Health Before and After Year 2000 in Ethiopia*, 19-20 October 1999, Addis Ababa, pp. 12-13.

[45] Cited in Save the Children (2001), *The Bitterest Pill: The Collapse of Africa's Systems*, London: Save the Children, p. 6; S. F. Murray (1966), 'The Costs of 'Adjustment': User Charges of Maternity Care,' in S.F. Murray (ed.), *Midwives and Safe Motherhood, International Perspectives on Midwifery*, Volume 1; Steven Russell (1996), 'Ability to pay for health care: concepts and evidence,' *Health Policy and Planning*, Vol. 11, No. 3; Steven Russell and Lucy Gilson (1997), 'User Fee Policies to Promote Health Service

Access for the Poor: A Wolf in Sheep's Clothing,' *International Journal of Health Services*, Vol. 27, No.2; A. Creese and J. Kutzin (1996), 'Lessons from cost recovery in health,' in *Marketing Education and Health in Developing Countries*, C. Colclough (ed.), Oxford: Clarendon Press.

# Chapter 7

# Summary and Conclusion

As population and urbanization increase, the federal government and municipal authorities in Ethiopia face a huge challenge in providing access to safe, reasonably priced drinking water, housing, and sanitation services for their citizens. At present the mismatch between the rapid growth of urban areas and the attendants for services is severe. Investments in urban infrastructure lag far behind the increase in urban population. Urban municipalities cannot provide adequate urban services or guide urban growth and expansion; nor can they offer accountability to their tax-paying residents. Over one and a half million urban people lack safe water and almost 3 million lack adequate sanitation. The liquid waste of over 8 million people is discharged with little or no treatment, and the solid waste of 5 million people is thrown away haphazardly. Three-quarters of urban dwellers live in unhealthy homes and neighborhoods. Congestion and industrial and vehicular emissions are on the increase. The resulting environmental ills pose excessive health hazards for the growing urban population.

Effective environmental health services can reduce many of the ecological problems that threaten health in urban areas. There are several steps that may be taken to minimize the environmental health burdens in urban areas, especially in Addis Ababa.

*Urban Housing* Public health problems in urban areas can generally be overcome through improvements in water facilities that generate reliable and accessible water supplies in good quality and sufficient quantity and through improvements in human waste disposal that lead to better sanitation. However, as fundamental as water supply and sanitation facilities are, they are not the only solutions to improve public health in the urban environment. Good health also requires decent dwellings. Housing provision constitutes a vexing problem in almost all urban areas in the country.

There are many barriers to the development of the urban housing sector. Government control of much of the improved and unimproved urban real estate, ineffective housing policies and regulations, and lack of infrastructure and financial resources have stunted urban housing development. Inflated urban land prices, intensified by housing needs and speculation, have condemned the majority of the urban population to poor quality dwellings. The problem has been compounded by the fact that housing is not affordable to most urban populations as a result of high construction costs, low household incomes, and a scarcity of financial resources. At present only small upper- and middle-income urban citizens have the income, job security, credit worthiness, and political connection to build homes or invest in urban real estate.

The urban land lease system has exacerbated the housing problem. Urban land is disposed of only on a lease basis. There is a huge gap between affordability and lease costs. This has consequently resulted in the growing of illegal settlements. Illegal settlements are basically encroachments on city-owned land by poor people. Because leasing urban land is not affordable for large segments of the urban population, many poor people resort to accessing land or housing outside the institutional arrangements. The problem will certainly get worse, because housing cannot keep up with the growing urban population. As the urban population and poverty levels increase, people will have no alternatives but to join the ranks of the urban homeless, to settle in unplanned, illegal settlements on the outskirt of cities/towns, or to move in with relatives who live in crowded living space.

The government cannot remain reluctant to confront the urban housing problem. The status quo is untenable because deteriorating housing conditions threaten to decrease current residential housing stocks, as municipal governments become less willing and able to maintain publicly-owned properties. Most apartment buildings and residential units are crumbling because of years of decay resulting from lack of repair and maintenance. The housing shortage in Addis Ababa is estimated at a quarter of a million units. Therefore, it is time for the government to find ways to denationalize urban real estates in favor of the private sector. The government should consider returning Kebele-controlled houses to their original owners or to their heirs. Originally, small property owners owned these houses. Prior to the nationalization of urban real estates in the mid-1970s, small-scale urban landlords supplied the majority of the low-income housing needs since more affluent landlords invested primarily in office buildings, sumptuous private houses, and expensive apartments that only expatriate residents and high-income groups could pay for (Kebbede, 1992: 50). The government should also seek ways to privatize properties now controlled by the government Agency for the Administration of Rental Houses (AARH). Most of these properties are now deteriorating for lack of proper maintenance. Returning Kebele-controlled urban properties to their former owners and disposing of properties managed by the AARH will have at least two positive impacts on the housing sector. First, it will improve the housing conditions, as owners will be willing to invest in rehabilitation. Second, with appropriate incentives owners will be willing to invest their extra income in rental houses for low-income groups, thereby alleviating the problem of housing shortages. However, following the denationalization of urban housing the government has to put in place laws that regulate rents to protect tenants from exorbitant rental hikes or forced eviction by unscrupulous landlords.

Municipalities should improve residential land development by increasing the distribution of serviced residential land to meet current and future demands. The government should also provide long-term mortgage loans to help low-income households afford homes by setting interest rates in accordance with their income levels. Affordability of housing can also be increased by allowing greater flexibility in the construction of homes that allow builders to accommodate the financial capacity of individual households (Gebre Giorgis, 2000: 16-28). The promotion of community-run savings and credit schemes can also enable low-income households to afford housing. Credit allows low-income households to make capital investments while saving schemes can promote housing improvement activities.

Home ownership can be an important investment. Privately owned homes are not left to deteriorate; residents have the incentive to make improvements and hence the value of these homes continues to increase. Income generated form housing investment stimulates economic growth. Housing construction requires materials, skilled labor, and transportation. All this will have multiple effects in the urban economy. With a legal title to a home, individuals can also gain access to loans and credits to start a new business or to expand an existing one. Taxes collected from all these multiplying economic activities will help urban municipalities to generate additional revenues and maintain infrastructure and upgrade and provide social services.

A secure home ownership can also be the catalyst to the reduction of poverty. Home ownership inspires people to invest in the home. To the poor, the ownership of a house means two things: first, the owner of a house does not risk eviction; second, the money that a household spends on its house translates automatically into assets. Value accrues in the house over time. Moreover, ownership of a house connects the household to infrastructure and services. The improvement in housing and basic infrastructure relieves the poor of ill health linked to destitute living conditions; and good health, in turn, reduces the amount of money one spends on medicines and healthcare and improves their quality of life (Satterwaite, 2002: 257-69). Quality housing can improve people's physical and mental health, increase work productivity, and create greater employment opportunities (Tulchin, 1986). Clearly, ownership of urban dwellings can lead to substantial environmental improvements in and around the household, and, in turn, stimulate good health. Therefore, there should be a move away from the centralized state-dominated approach to urban land allocation and use.

Real estate taxes are an important source of municipal revenue for the construction and maintenance of infrastructure and for the provision of services. At present urban areas generate little revenue from property tax because the government controls most of the real estates—land and housing. Most of the revenue from government-owned housing rental is used up by administrative or bureaucratic outlays and little is left for improving urban infrastructure and services. Current rental charges for publicly owned properties are far below the market price. If urban real estates are privately owned rather than government owned, it would generate substantial revenue for municipal governments. For example, the only taxable properties in Addis Ababa are those that are privately owned which account for only 30 percent of the housing stocks in the city (Ayenew, 1999: 28). Privatizing of government owned buildings and rented houses, which now accounts for 40 percent of the city's real estates (Ayenew, 1999: 10), would have more than doubled Addis Ababa's revenue from property taxes. Such additional revenues could have helped the municipality to become financially solvent and to discharge its duties as a provider of basic urban services and an effective manager of the urban environment. This, of course, assumes the existence of an efficient municipal tax collection system in place, which is not the case. For the most part, tax collections in urban areas are irregular and grossly inadequate, one of several reasons why urban municipalities' coffers are minuscule. There is a need for revising old tax codes, introducing new taxes, better administration of taxes, and increased user fees and service charges that now fall way below the actual costs of providing services.

*Healthcare Services*   Healthcare facilities in Ethiopia are poorly equipped, inadequate, lack trained staff, and are unevenly distributed. The health sector needs more than the current financial resources in order to provide a minimal level of healthcare. It is worthwhile for public health policy-makers to look into the experiences of other countries that have achieved better health for most or all of their citizens despite being poor. Mehrotra and Jolly (1998)[1] studied several low-income countries that achieved high levels of life expectancy and lower levels of child mortality. They found that these countries shared a number of policy measures: provision of universal health services for all citizens paid out of government revenues with resources allocated towards the lower end of the pyramid; higher level of budgetary expenditure on health services, more equitably distributed; provision of strong incentives that encouraged health personnel to serve in small towns and rural areas; emphasis on training nurses and village health workers/community health workers/primary health midwives; provision of comprehensive and widespread maternal and child health services at the primary level; an increase in the proportion of births attended by trained health personnel and good postnatal follow-up visits; effective and timely referral systems; provision of widespread immunization services; a supportive and complementary role on the part of the private sector; provision of a nutritional floor through targeted subsidies to households facing food insecurity; improved water supply and sanitation facilities; and promotion of health and hygiene education.

    One must recognize that many factors contribute to ill health. Poverty causes many of the country's health problems. The promotion of good nutrition will have very little impact when the cause of malnutrition is privation. The health sector alone cannot solve or alleviate such a problem; improvements in the agricultural sector must be sought as well in this situation. Such improvements may include appropriate agricultural policies relating to, for instance, land ownership, input provision, and food prices. Improvements in water supplies and sanitation will bring a more long-lasting solution for children suffering from gastroenteritis, not in a mere use of oral dehydration solutions. Water and sanitation sectors ought to play a major role in this regard, for many diseases that affect the health of most Ethiopians stem from the interrelated problems of water quality, availability, sanitation, and hygiene. Health education activities at healthcare facilities are important: however, the involvement of the education sector in such efforts would have far greater impacts.

The MOH should not be left with the sole responsibility for healthcare. The country can achieve sustained improvement in health only if each of the public sectors—including, the agriculture, education, water, sanitation, and housing—makes an appropriate contribution. For this to happen effectively, a mechanism for inter-sectoral coordination and cooperation must be created and this is where the MOH can play an effective leading role. The government realizes and preaches about the need for inter-sectoral cooperation, but it has not been able to translate its rhetoric into deeds. Poor health is detrimental to economic development due to associated losses in human lives and productivity, and because of resources that need to be channeled into medical care. Good health and all its related services are imperative to sustainable development, and thus must be given an utmost priority in terms of resource allocation.

*Water supply*  Water is a vital resource and Ethiopia is fortunate to possess it in abundance. It has an estimated 112 billion cubic meters per year, including 2.6 billion cubic meters of underground water. However, the analysis of water demand and supply reflects that a huge gap exists despite availability of adequate water resources in the country, and this gap is bound to rise in the future. Other aspects of concern in the water sector are the poor quality of the water supply and inefficient services and inequities in water supply levels. Unaccounted-for water losses in the urban distribution systems are massive. In some urban areas, up to one-half of the water fails to reach the end user. Most of this water is lost through leaking pipes, overflowing service reservoirs after abstraction, pumping or treating, or during distribution. This is a huge loss of precious resource. It is also a loss of revenue for municipalities as well. This revenue, if collected, could have been used for maintenance and to build new facilities to meet the needs of more urban households. The high volume of water loss is due largely to inadequate maintenance through the years. This problem begs for concerted efforts on the part of municipal water agencies and the public to put more emphasis on water management and conservation and on the prevention, detection, and repair of leaking pipes. Upgrading water distribution systems is extremely important to reduce the huge amount of water lost through leakage. Total elimination of water losses from systems is, of course, impossible. According to water experts, system losses of 10 percent or less are considered good enough.

Most municipal water agencies' efforts are overwhelmingly devoted to meet the growing water demands. Supplying urban residents with adequate potable water is their overriding task. Owing to this, the management side of water is under-emphasized. However, the savings that can be made from proper water demand management are quite high. Water demand management involves the management of water from point of abstraction (reservoir, for example) to point of sale. According to the United Nations Habitat, typical water demand management activities may involve: 'Efficient operation of the water supply system from abstraction point to consumer; pressure management; night flow monitoring and water audit; metering (bulk and domestic); removing of wasteful devices and installation of water saving devices; tariff structures (for instance, instituting a pricing system which charges a higher rate as a higher volume of water is consumed) and billing procedures; training and capacity building; active and passive leakage control; general education and public awareness; school education; water-wise gardening; use of rainwater tanks;' and recycling of sullage and wastewater for non-potable use (AAWSA/UNCHS, 2000: 9-10).

All these activities go hand in hand with effective water resource management. Water resource management involves proper and effective management of catchments to abstract water from a dam, river, or aquifer system into supply system. Like water demand management, water resource management involves wide ranging management activities, including 'efficient operation of surface and groundwater resources; equitable allocation of the resource to the different user sectors; management of stream flow reduction activities; management of land use to minimize soil erosion; conservation and rehabilitation of reservoir catchments; education and public involvement; institutional development and capacity building; and a management of the water quality' (AAWSA/UNCHS, 2000: 9-10). It also involves avoiding mega-engineering solutions like dams and opting for micro-dams that can be constructed, managed, and operated with local labor and finance.

Many argue that because public water agencies are inefficient and incapable of raising the capital needed to expand and improve water provision, private enterprises should be allowed to manage water systems to help overcome safe drinking water shortages (Tagliabue, 2002).[2] The rationale is that private companies can more effectively finance and deliver safe water. However, making water, a resource necessary to life, a profit-making enterprise may not produce a cure for poorly performing public water systems. This is due largely to the fact that private firms invest in water sector only if consumers pay tariffs sufficient enough to generate high returns; they are in it for profit, not for ensuring social justice. Allowing the private sector to manage water systems may thus expose poor households to exorbitant water tariffs or force them to resort to accessing water from polluted streams and rivers.[3] This may, in turn, increase the risk of water-borne diseases, including parasitic diseases and cholera. Privatizing water has often led to unhappy consequences. In South Africa, for example, water privatization measures resulted in water supply cuts for people who were too poor to pay their bills, and this led to the outbreak of a cholera epidemic in KwaZulu-Natal. The government there 'admitted the poor had little access to safe drinking water because of the prohibitive costs' (Nyambura, 2003). Similarly, the comeback of water-borne guinea worm in a region in Ghana has been attributed to 'unaffordable capital contributions from local communities' demanded by a

World Bank water and sanitation project.[4] Others also argue that once such a vital resource is handed to a private corporation, especially to a predatory and powerful entrepreneurial transnational corporation, the corporation could subject water to the rules of international competition and the discipline of the World Trade Organization,[5] thus reducing or even completely removing the government's ability to regulate the service on behalf of its citizens.[6] Additionally, privatizing such an essential service as water delivery may lead to social upheaval stemming from public protests against the imposition of prohibitive fees for drinking water, as happened in developing nations such as Bolivia, Paraguay, the Philippines, and South Africa (Gleick, et. al., 2003: 58; Klare, 2001: 25). Even the most developed nations such as the United States are not excepted from protests against the sale of public water services to multinational corporations. The city of Atlanta, Georgia, for example, recently restored its water system to public control and management after four years of privatization, citing price hiking and poor services.[7] The performance of major global companies currently delivering private water and wastewater services both in the developed and the developing countries has been well documented: 'huge profits, high prices for water, cut-offs to customers who cannot pay, little transparency in their dealings, reduced water quality, bribery and corruption.'[8]

        In Ethiopia water resource management has not focused on cost recovery or economic viability. The sector has existed on government, foreign, and NGO aid. As a consequence the sector is economically defunct and non-sustainable. At present all municipal water agencies subsidize water to make it affordable for poor households, but the subsidies actually benefit those that are better off. Existing water tariffs in the vast majority of urban areas do not cover even the operating and maintenance costs, let alone the costs incurred to underwrite the capital investment—making it exceedingly difficult to operate a self-financing viable water supply system. Adequate pricing of water, efficient billing and bill collection systems, and repairing leaks can substantially improve the financial viability of the existing systems. Thus, increasing user fees are unavoidable because otherwise there would be no resources for public water agencies to invest in water systems or maintain and operate them adequately. The need for more investment, better maintenance, and good operation is quite obvious. That means adequate user fees that consumers can afford do not present the only solution.[9] The introduction of improved pricing structures that reflect the real cost of water supply also promotes efficient use of water.

        From the perspective of societal welfare the government has a responsibility to provide its citizens with adequate, safe drinking water. Government should maintain public ownership of water resources. However, community participation in all stages of water resource management (planning, design, operation, maintenance, and assessment) is of paramount importance in the success of water supply development programs. Community participation provides a sense of ownership and responsibility. There are many examples of successful community-involved self-help water supply schemes elsewhere in

Africa that are worth looking into in order to promote sustainable water systems in Ethiopia. Colin Glennie's (1983: x) excellent study of the self-help water supply scheme in Malawi provides a number of prerequisites to success. These preconditions may include: reliance on local initiative and involvement of the community (especially women) throughout the planning process; use of a local committee structure which places responsibility for the project on those who are to benefit most; integration of the water project into local government institutions; the creation of cohesive inspired teams of field workers selected from the local communities and given focused in-service training; putting in place well-designed maintenance programs that rely on the community and its assumption of long-term responsibility for the water supply installation; and finally pungent central government commitment to the well being of its citizens.

*Pollution Control* Although industry brings economic development to urban centers, urban communities, especially those in poverty, pay a high price for that development in their declining health and in the health damage to their surrounding environments. At present nearly all industries freely dispose of byproducts in the air and water without concern for public or environmental health: their concern is purely economical. Localized air pollution is a major problem for people living around industries. River pollution from Addis Ababa-based industries and untreated sewage poses health problems in downstream rural settlements. The EPA needs to take the lead in developing and enforcing strict environmental laws that make industries responsible for their impacts on the environment. The EPA should require all industries to treat their wastes before they release them into the environment. Taxation of emissions would effectively limit pollution. In addition, industries that go above and beyond the required environmental improvements could receive tax credits, and thus environmental health could possess an economic incentive. At any rate, a strong EPA is needed to enforce existing laws, to formulate new ones, and to initiate creative solutions to urban environmental problems.

*Waste Management* Urban areas produce increasing amounts of solid and liquid wastes. The Addis Ababa municipality generates about 2,000 cubic meters of solid waste per day of which only 60 percent finds its way in the municipal dumpsite. The municipal solid waste disposal site has outgrown its service but the city continues to use it. This poorly designed and ill-managed dumpsite has contaminated the soil, groundwater sources, and the nearby residential and farm environments.[10] Much of the hazardous wastes generated from large and small-scale industrial enterprises invariably find their place in the city's waterways and open spaces. Large quantities of infectious biomedical wastes also end up in waterways and in municipal waste dumps—without treatment—due to the absence of dedicated disposal facilities. The potential risks to health and the environment from improper handling and disposal of solid wastes are high.

Liquid waste disposal management in urban areas is also pathetically deprived. The sources of liquid wastes include households, public and commercial establishments, and a wide range of industrial establishments. Much of the liquid wastes generated in urban areas are often improperly discharged in open spaces and into waterways. Human excreta, one of the most common types of urban liquid waste, are a widespread source of pollution in urban areas. Only less than one-half of the urban population is served with sanitation services and, as a consequence, defecation in open spaces, drainage channels, and river ravines is the most common practice. Urban areas in Ethiopia have no sewage system at all, if we discount the existence of a sewage system in Addis Ababa that serves only small fractions of the city's households and commercial establishments. The pervasiveness of waterborne diseases ensuing from sewage, excreta, or fecal contamination begs the question of what can be done to address the problem of sanitation in urban areas.

It is clear that the production of liquid waste increases in direct proportion to the increase in urban water consumption. Most of the liquid waste that urban areas produce at present is either not treated at all or receives inadequate treatment. Hence, it is crucial that municipalities collect and treat adequately their liquid waste in order to reduce the environmental and health risks it poses for people living in and around urban centers. Sufficiently treated liquid waste could also be put to good uses such as for cleaning, washing, and cooling. Individual households and neighborhoods cannot be expected to dispose of liquid wastes appropriately and at a great enough distance from their residences. It is the role of municipal governments to provide sewage networks that can carry waste safely to appropriate disposal sites and properly treat them.

Urban municipalities alone are now responsible for providing all urban services. As has been shown, none of the urban governments in the country are able to provide the full scope of urban amenities. There are too many unmet needs at present and the problem has become too heavy for urban municipal governments to handle. Consequently, it may be necessary to consider the involvement of local private entrepreneurs in the provision of some urban services. Allowing the private sector to provide urban environmental services, especially liquid wastewater treatment and solid waste collection and disposal, can offer potential cost savings for urban municipalities. Local private enterprises are already filling important gaps in the provision of urban services (for example, urban transportation, water vending, and sanitation). The private sector has done a good job in providing sanitation service as can be witnessed by the operational success of the few sludge hauling companies that exist in the capital. It is also time to encourage private sector involvement in operating public toilets and baths in cities and towns. People are forced to defecate in the open largely because public provision of sanitation facilities is either acutely inadequate or totally absent. Making privately run public toilets and baths as widely available as possible can significantly reduce environmental health risks. However, this move towards greater participation of the private sector requires municipal supervision.

Municipal governments must develop and enforce effective regulations for privately run services in order to maintain high quality services at reasonable prices and to ensure free and fair competition. Providing the public with accurate information on service quality, service fee regulations, and regular supervisions should be the responsibility of municipal governments. Additionally, municipal governments should regulate the environmental impacts of private activities to ensure that such activities do not threaten the health and safety of the general population. Of course, an effective supervisory role demands the institution of competent and trustworthy municipal governments that are accountable to their residents.

It is obvious that urban areas have a serious waste disposal problem. It is also quite apparent that municipal governments alone cannot solve this problem—like so many other urban problems. Improvement in the quality of the urban environment cannot be achieved without the active involvement of urban residents. Municipal administrations should actively encourage residents to organize by providing them opportunities to share responsibilities in the management and development of their city or town. Organized citizen groups can tackle the environmental problems that menace their neighborhoods. Neighborhood-organized groups can involve themselves in sanitation management, including garbage collection, sewage disposal, cleaning of streets, open neighborhood markets, storm drains, maintenance of recreational areas and green spaces, planting trees, and flowering plants. For neighborhoods where the environmental degradation is too big for local groups to tackle, municipal governments should provide funding, tools, and guidance to residents to upgrade and maintain their environment (Simioni, 1999). Citizen participation not only improves the quality of the urban environment but also reduces the cost of environmental protection and management for municipal governments (Ling, 1999).

Thus, municipal governments must actively encourage and support communities and grassroots organizations. They should bolster people's capacity and ingenuity to do things for themselves and their communities. Citizens' awareness about the health and environmental importance of proper waste disposal must be heightened via public education campaigns using the mass media, community associations, neighborhood institutions, and youth organizations. Waste management authorities should develop educational materials for urban communities and schools. Such materials should educate the public and the youth about waste reduction, reuse, recycling, and disposal. The process of involving communities in the betterment of their living conditions is fundamental to sustainable development.

The most successful experiences that have addressed environmental problems associated with urban development elsewhere in the developing economies 'have been associated with politicians and civil servants with strong commitments to democratic practices, greater accountability to citizens and partnerships with community-based organizations and non-governmental organizations.'[11] The democratization of urban governance will 'require extraordinary political will and intense pressure from local business interests, non-governmental organizations, and citizens' alliances ' (El-Shaks, 1997: 513). One would imagine that the current crises in urban governance and in the delivery of built environment services and the general degradation of the quality of the country's urban environments might help induce such actions.

Many of the urban problems in Ethiopia stem from the lack of good governance that empowers people to have control over their lives. The government claims that its decentralization, democratization, and capacity-building efforts have empowered people. But this claim does not correspond to reality.[12] Empowerment does not mean forcing people to participate in superficial discussions after decisions on critical social and economic issues have already been made at the top. Rather, it means augmenting the capability of citizens 'to influence the state institutions that affect their lives, by strengthening their participation in political processes and local decision-making. And it means removing barriers—political, legal, and social—that work against particular groups and building the assets of [ordinary] people to enable them to engage effectively in' development (World Bank, 2000c: 39). Participation ensures that the choices and ideals of citizens or communities are reflected in the preference and design of their development programs; participation helps monitor implementations, transparency, and accountability; and, above all, it gives citizens a say over their lives (Ibid., 88). Unfortunately, this is not what is happening in Ethiopia, genuinely speaking. People still lack real voice, power, and representation. The intent of the party in power is to hold onto the reins and continue to manage citizens participation on its own terms. Building the institutions of democracy is not an easy task, especially in a country with virtually no such legacy. However, Ethiopia deserves a more accountable, transparent, and democratizing political regime.

## Notes

[1] Cited in Save the Children (2001), *The Bitterest Pill of All: The Collapse of Africa's Health Systems*, Save the Children, London, p. 17.

[2] The World Bank and the IMF are widely promoting privatization of water across the developing countries in spite of considerable opposition. Water sector development loans to poor countries are mostly conditioned on water privatization and full cost recovery.

[3] Evidence shows that when the water supply is privatized the urban poor will either not be covered by such a service or will have to pay fees that often exceed those paid by higher income residents. Cost increases can make water less accessible and less affordable to poor people that make up the majority of the country's population. When that happens the burden falls on women and children who must travel farther and toil harder to fetch water—almost invariably resorting to unsafe water sources. See Trudy Harpham and

Emma Grant (2002), 'Health, Health Services and Environmental Health,' in *Urban Livelihoods: A People Centered Approach to Reducing Poverty*, Carole Rakodi and Tony Lloyd-Jones (eds.), London: Earthscan, pp. 165-179; Jorge E. Hardoy and David Satterthwaite (1992), *Environmental Problems in Third World Cities*, London: Earthscan.

[4] Rick Rowden, 'World Bank Water Policies Undermine Public Health.' Available at: www.resultsusa.org.

[5] In January 2003, Ethiopia submitted the preliminary application to join the World Trade Organization.

[6] Developing nations are increasingly under pressure from international development agencies to privatize their water supplies. The World Bank, the IMF, and the WTO are powerful global institutions that promote the privatization of water. Of the 276 water supply project loans provided by the World Bank between 1990 and 2002, one-third of the projects required borrowing countries to privatize their waters in some form before they received funds (International Consortium of Investigative Journalists, 'Promoting Privatization.' Available at:
http://www.icij.org/dtaweb/water/default.aspx?SECTION=ARTI). The World Bank and the IMF also demand water privatization or full cost recovery in exchange for debt relief. The WTO is mandated to remove tariff and non-tariff barriers to the free flow of goods, including water, across national borders and is negotiating free trade in water services through the General Agreement on Trade in Services (GATS).' (National Non-Profit Public Interest Organization, 'The 3rd World Water Forum: A Civil Society Backgrounder.' Available at: http://www.citizen.or/cmep/water/cmep_Water/articles.cfm?ID=9130). It is an irony that while developing countries are pushed to privatize their water services, water services are still managed by public agencies in most developed nations. In the United States, for example, 85 percent of the people obtain their water from public agencies (Jon Luoma 2002), 'Contamination, riots, rate of increases, scandals: From Atlanta to Manila, cities are confronting the true cost of water privatization,' *MotherJones*, November and November, p. 35; pp. 35-38 and 88).

[7] Prabha Khosla and Rebecca Pearl, 'Gender Differences in Water Use and Management,' *Women's Environment and Development Organization*. Available at: http://www.wedo.org/sus_dev/untapped3.htm.

[8] Maude Barlow, 'The 3rd World Water Forum: A Civil Society Backgrounder,' National Non-Profit Public Interest Group (USA), March 2003. Available at: http://www.citizen.org/cmep/Water.cmep_Water/articles.cfm?ID=9130.

[9] Most existing water tariffs are not only low but also are uniform. A uniform tariff rate per cubic meters tends to be insensitive to the ability of customers to pay. Low-income groups are usually denied access to services and high-income population that can pay more are not potentially tapped for revenue. Water experts recommend block or stepped tariff rates to resolve this inequity. Block rates are dependent on the volume of consumption and, thus, have varying price per cubic meter. This system is considered more appropriate for the Ethiopian water system than the uniform rate. These rates allow major users of water and

those who can afford high use to pay more. In contrast, small users of water pay less (Ministry of Water Resource (1998d), *Water Supply Development and Rehabilitation: Tariff and Asset Valuation Study*, Addis Ababa: MOWR).

[10] Urban areas, especially Addis Ababa and the larger ones, need to build disposal facilities that are designed to permanently isolate solid wastes from the environment (away from settlements and water resources). This requires burial of the wastes in a landfill. If waste is tipped into a hole in the ground and covered with soil, it will ultimately turn into harmless substance through natural chemical and biological processes. The problem of leachate seeping into the groundwater system or surface water can be avoided—though not totally—by lining the landfill with non-porous clay and/or synthetic material (K.A. Gourlay (1992), *World of Waste: Dilemmas of Industrial Development*, London: Zed Press, pp. 91-92).

[11] Gordon McGranahan, Liliana Miranda, David Satterthwaite, and Luz Stella Velasquez (2002), 'Striving for Good Governance in Urban Areas: The Role of Local Agenda 21s in Africa, Asia and Latin America,' International Institute for Environment and Development, May. Available at: http://www.iied.org/pdf/wssd_08_la21s.

[12] It is true that the government has made significant attempts at decentralization. However, these efforts have proved ineffective because regional governments still have limited power in decision-making and allocation of resources. Moreover, development programs and projects are formulated and implemented with little or no input from the affected stakeholders.

# Bibliography

Aberra, Edlam (2001), 'Air Pollution in Addis Ababa: The Predicament of Chile Sefer' in
    *Local Environmental Change and Society in Africa*, M.A. Mohamed    Salih
(ed.),    Boston: Kluwer Academic Publishers: 177-202.
Adam, Adinew and Mohammed, Nuri (1997), *Water Supply in Addis Ababa*, Central
    Laboratory Service, Addis Ababa: Addis Ababa Water and Sewage    Authority.
Adams, John (1999), *Managing Water Supply and Sanitation in Emergencies*, Oxford:
    Oxfam. *Addis Tribune* (1996), 'Egypt Waters Her Desert With the Nile
    (What is  Ethiopia doing with the River?),' 30 September.
Addis Ababa City Administration (2000), *Proceedings of the Workshop on the State of
    Industrial Pollution in Addis Ababa and Possible Remedies*,    Addis    Ababa:
    Environmental Protection Bureau, Addis Ababa City Administration.
Addis Ababa City Government (2001), *Proceeding of the Workshop on Sanitation
    and Beautification of Addis Ababa*, Addis Ababa: Environmental Protection
    Bureau, Addis Ababa City Government.
_____ . (1999), *Proceedings of the Conference on the Addis Ababa City
    Administration Conservation Strategy*, 30 June-2 July, Addis Ababa:
    Environmental Protection Bureau, Addis Ababa City Government.
Addis Ababa Master Plan Revision Office (1999a), *Land Use and City Structure
    Studies of Addis Ababa and the Metropolitan Area*, Addis Ababa, Addis Ababa
    Master Plan Revision Office.
_____ . (1999b), *Addis Ababa and its Planning Practice*, Addis Ababa, Addis Ababa
    Master Plan Revision Office.
_____ . (2000a), *Addis Ababa Revised Master Plan Proposals: Draft Summary*,
    Addis Ababa: Addis Ababa Master Plan Revision Office.
_____ . (2000b), *Environmental Profile and Environmental Management of Addis
    Ababa*, Addis Ababa: Addis Ababa Master Plan Revision Office.
_____ . (2001), *Housing Component: Improvement and Development Strategy*,
    Addis Ababa, Addis Ababa Master Plan Revision Office.
Addis Ababa Water and Sewerage Authority (1957), *Water Supply and Sewage Services in
    Addis Ababa*, Addis Ababa: Addis Ababa Water and Sanitation Authority.
_____ . (1972), *Water Supply Services for the City of Addis Ababa* Addis Ababa:
    Addis Ababa Water and Sewerage Authority.
_____ . (1984), *Addis Ababa Water Resources Reconnaissance Study, Volume I:
    Main Report*, Addis Ababa, Addis Ababa Water and Sewerage Authority.
_____ . (1993), *Master Plan Study for the Development of Liquid Waste Facilities
    for the City of Addis Ababa: Existing Situation and Design Criteria Report,
    Volume 2*, Addis Ababa: Addis Ababa Water and Sewerage Authority.
_____ . (1994), *The Study on the Problem Analysis of Means to Control Water
    Quality Deterioration of Lagadadi Reservoir, Final Report, Volume 1: Main
    Report*, Addis Ababa: Addis Ababa Water and Sewerage Authority.
_____ . (1996), *Master Plan for the Development of Liquid Waste Facilities for the
    City of Addis Ababa*, Addis Ababa: Addis Ababa Water and Sewerage Authority.

_____ . (1996), *Master Plan for the Development of Liquid Waste Facilities for the City of Addis Ababa,* Addis Ababa: Addis Ababa Water and Sewerage Authority.

Addis Ababa Water and Sewerage Authority/United Nations Center for Human Settlement (2000), *Managing Water for African Cities: The City of Addis Ababa, Water Demand Management Plan,* Addis Ababa: Addis Ababa Water and Sewage Authority.

Ahderom, Techeste (1987), 'Basic Planning Principles and Objectives Taken in the Preparation of the Addis Ababa Master Plan, Past and Present' in *Proceedings of the International Symposium on the Centenary of Addis Ababa,* November 24-25, 1986, Ahmed Zekaria, Bahru Zewde and Taddese Beyene (eds.), Addis Ababa: Institute of Ethiopian Studies, Addis Ababa University: 247-269.

Alemayehu, Tamiru (2001), 'The Impact of Uncontrolled Waste Disposal on Surface Water Quality in Addis Ababa, Ethiopia,' *SINET: Ethiopian Journal of Science,* 24, 1: 93-104.

Allan, J. Anthony (1992), 'The Changing Geography of the Lower Nile,' in *The Changing Geography of Africa and the Middle East,* Graham P. Chapman and Kathleen M. Baker (eds.) New York: Routledge: 165-190.

Altman, Lawrence K. (2002), 'AIDS in 5 Nations Called Security Threat,' *The New York Times,* 1 October.

Amha, M., et. al. (1995), 'Child health problem in Ethiopia,' *Ethiopian Journal of health Development,* 9: 167-185.

Ankrah, E. M. (1991), 'AIDS and the social side of health,' *Social Science and Medicine,* 32: 967-980.

Assefa, Asnakew (1998), *Resettlement as a Measure of Upgrading Slums in Addis Ababa: A Case Study of 300 Relocated Housing Units,* paper prepared for a workshop on 'Housing Delivery and Project Management,' Addis Ababa, IHS.

Assen, Eshetu (1987), 'The Growth of Municipal Administration and Some Aspects of Daily Life in Addis Ababa, 1910-1930,' in *Proceedings of the International Symposium on the Centenary of Addis Ababa,* November 24-25, 1986, Ahmed Zekaria, Bahru Zewde and Taddese Beyene (eds.), Addis Ababa: Institute of Ethiopian Studies, Addis Ababa University: 79-95.

Ayehunie, Seyou, M.D., (2001), 'Africa AIDS strain not being targeted in vaccine trials,' *EuropaWorld,* 16 March.

Ayenew, Meheret (1999), *The City of Addis Ababa: Policy Options for the Governance and Management of a City with Multiple Identity,* Forum for Social Sciences, Discussion Paper No. 2.

Baker, Jonathan Baker (1994), 'Small Urban Centers and their Role in Rural Restructuring,' in *Ethiopia in Change: Peasant, Nationalism and Democracy,* Abebe Zegeye and Siegfried Pauswang (eds,) London: British Academic Press: 152-171.

Baross, P. and Linden, J. Van Der (eds.) (1990), *The Transformation of LandSupply Systems in Third World Cities,* Aldershot: Avebury-Gower.

Befekadu, Sisay (1999), 'Fountains of Explosives,' *The Reporter,* 14 April.

Bekele, Seble (1999), 'The Challenges of Cleaning Addis Ababa,' *Addis Tribune,* 4 August.

Bekerie, Ayele (1997), *Ethiopia: An African Writing System,* Lawrenceville, N.J.: Red Sea Press.

Berhe, Tesfaye, Ghirmay, Zereu, and Abdella, Said (2000), *Assessment of Little Akaki River Water Pollution, Addis Ababa*, Addis Ababa: Environmental Protection Bureau, Addis Ababa City Government.

Beyene, Atnafe (1997), 'The Role of Community Participation in the Sustainable Development of Water Resources Management,' *Water*, 1, 1 (June): 33-38

Bezabih, Werotaw (2000), *Economic Significance of Addis Ababa: Assessment of the Addis Ababa Urban Economy*, Addis Ababa: Department of Economics, Unity College.

Bhalla, Nita (2001), 'War Devastated Ethiopian Economy,' *BBC News Service*, 7 August.

Black, Robert E., M.D. (1999), *Reducing Perinatal and Neonatal Mortality*, Report on a Meeting in Baltimore, Maryland, 10-12 May, volume 3, No. 1.

Bunce, Nigel J. (1994), *Environmental Chemistry*, Winnipeg, Wuerz Publishing. *Business Review* (2002), 'Association Calls for Inclusion of Provision in Criminal Code Against Violence on Women,' 1, 45 (4 April).

Butzer, Karl W.(1981), 'Rise and Fall of Axum, Ethiopia: A Geo-Archaeological Interpretation,' *American Antiquity*, 46: 472-495. Buschkens, W. F. L. and Slikkerveer, L. J. (1982), *Health Care in East Africa: Illness Behavior of the Eastern Oromo in Hararghe (Ethiopia)*, Van Gorcum Assen, The Netherlands.

Cairncross, Sandy and Feachem (1993), Richard G. *Environmental Health Engineering in the Tropics: An Introductory Text*, New York: Wiley.

Central Statistical Office (1967), *Statistical Bulletin*, No. 1, Addis Ababa: CSO.

_____ . (1984), *The 1984 Population and Housing Census*, Addis Ababa: Central Statistical Office.

Central Statistical Authority (1991), *The 1990 Fertility and Family Survey*, Addis Ababa: CSA.

_____ . (1994a), The 1994 Population and Housing Census, Addis Ababa: Central Statistical Authority.

_____ . (1994b), The 1994 Population and Housing Census of Ethiopia: Results at Country Level, Volume II Analytical Report, Addis Ababa: Central Statistical Authority.

_____ . (1995), *The 1994 Population and Housing Census Result for Addis Ababa*, Addis Ababa: Central Statistical Authority.

_____ . (1997), *Report on Urban Informal Sector Survey*, Addis Ababa: Central Statistical Authority.

_____ . (1999), *Report on the 1998 Welfare Monitoring Survey*, Statistical Bulletin No. 224, Addis Ababa: Central Statistical Authority.

_____ . (2001a), *Demographic and Health Survey 2000*, Addis Ababa: Central Statistical Authority.

_____ . (2001b), *Statistical Abstract 2000*, Addis Ababa: Central Statistical Authority.

Chapman, Stephen R. (1998), *Environmental Law and Policy*, Columbus: Prentice Hall. Chapple, D. (1987) 'Some Remarks on the Addis Ababa Food Market Up To 1935,' in *Proceedings of the International Symposium on the Centenary of Addis Ababa*, November 24-25, 1986, Ahmed Zekaria, Bahru Zewde and Taddese Beyene (eds.), Addis Ababa: Institute of Ethiopian Studies, Addis Ababa University: 143-160.

Chen, B. H., Hong, C. J. and Smith, K. R. (1990), 'Indoor air pollution in developing countries,' *World Health Statistics Quarterly*, 43: 127-138.

Conway, Declan and Hulme, Mike (1996), 'The Impacts of Climate Variability and Future Climate Change in the Nile Basin on Water Resources in Egypt,' *Water Resources Development*, 12, 3: 277-296.

Council of Representatives, Proclamation No. 80 (1993), Proclamation to Provide for the Lease Holding of Urban Lands, *Negarit Gazeta*, 53rd year, No. 40, 23 December.

Council of Representatives, Region 14 Administrative Regulations No. 4 (1994), Urban Lands Lease Holding Regulations, *Addis Negarit Gazeta*, 3rd year, No. 1, 3 November.

Crewe, Emma (1995), 'Indoor Air Pollution, Household Health, and Appropriate Technology,' in *Down to Earth: Community Perspectives on Health, Development, and the Environment*, Bonnie Bradford and Margaret A. Gwynne (eds.), West Hartford: Kumarian Press: 92-99.

Crummey, Donald (1987), 'Some Precursors of Addis Ababa: Towns in Christian Ethiopia in the 18th and 19th Centuries,' *Proceedings of the International Symposium on the Century of Addis Ababa*, November, 24-25, 1986: 9-31.

Cunningham, William P. and Saigo, Barbara W. (2001), *Environmental Science: A Global Concern*, New York: McGraw Hill. *Daily Monitor* (Addis Ababa) (2002), 'About 250,000 children in Ethiopia live with HIV,' 6 February.

Desta, Asfaw and Meche, Hailu (1999), 'Review of the Development of Health Services in Ethiopia,' in Proceedings of the 10th Annual Scientific Conference of the Ethiopian Public Health Association: Health Before and After 2000 in Ethiopia, 19-20 October 1999: 13-18.

Department of Economics, Addis Ababa University (1995), *Report on the 1994 Socio-Economic Survey of Major Urban Centers in Ethiopia*, Addis Ababa University: Institute of Development Research.

Desta, Naomi (1996), *Land Management Reform in Addis Ababa, Ethiopia: Implementing A Public Leasehold System*, M.A. Thesis submitted to the Department of Urban Studies and Planning at Massachusetts Institute of Technology.

El-Shakhs, Salah (1997), 'Toward Appropriate Urban Development Policy in Emerging Mega-Cities in Africa,' in *The Urban Challenge in Africa: Growth and Management of Its Large Cities*, Carole Rakodi (ed.), Tokyo: United Nations University Press: 497-526.

Environmental Protection Authority (1997), *Preliminary Survey of Pollutant Load on Great Akaki, Little Akaki and Kebena Rivers*, Addis Ababa: Environmental Protection Authority.

_____ . (2001), *National Review Report on the Implementation of Agenda 21, A Report Prepared for the United Nations Conference on Environment and Development*, Johannesburg, South Africa, June 2002, Addis Ababa: Environmental Protection Authority.

Environmental Protection Authority and United Nations Industrial Development Organization (2001a), *Ecological Sustainable Industrial Development*, US/ETH/068 Ethiopia, Volume 1, Addis Ababa: Environmental Protection Authority.

_____ . (2001b), *Ecological Sustainable Industrial Development*, US/ETH/99/068 Ethiopia, Volume 2, Addis Ababa: Environmental Protection Authority.

_____ . (2001c), *Ecological Sustainable Industrial Development*, US/ETH/99/068 Ethiopia, Volume 3, Addis Ababa: Environmental Protection Authority.

Esrey, Steven A. (1996), 'Water, Waste, and Well-being: A Multi-country Study,' *American Journal of Epidemiology*, 143, 6: 608-622.

Ethiopian Economic Association/Ethiopian Economic Policy Research Institute, *Quarterly Report on the Macroeconomic Performance of the Ethiopian Economy*, Volume1. Number 1 (November 2003).

Ethiopian Health and Nutrition Research Institute (2001), *Annual Report*, Addis Ababa: Ethiopian Health and Nutrition Research Institute.

Ethiopian Valleys Development Studies Authority (1992), *A Review of Water Resources of Ethiopia*, Addis Ababa: EVDSA.

Ezra, Markos and Gebre Silassie, Seyoum (1998), *Handbook on Population and Family Life Education for Secondary School Teachers in Ethiopia*, Addis Ababa: Institute for Curriculum and Development Research.

Fainstein, Susan S. and Markusen A. (1993), 'The Urban Policy Challenge: Integrating Across Social and Economic Development Policy, *North Carolina Law Review*, (June): 1463-1486.

Family Health International (2000a), 'Many Youth Face Grim STD Risks,' *Network*, 20, 3: 4-9.

_____ . (2000b), 'Sex Education Helps Prepare Young Adults,' *Network*, 20, 3: 10-15.

Federal Democratic Republic of Ethiopia (1998), *Policy on HIV/AIDS of the Federal Democratic Republic of Ethiopia*, Addis Ababa: Master Printing Press.

_____ . (1995), *Ethiopia: National Program of Action for Children and Women, 1996-2000*, Addis Ababa: Federal Democratic Republic of Ethiopia.

_____ . (2000), Proclamation No. 200/2000: Public Health Proclamation, *Federal Negarit Gazeta*, No. 28, Addis Ababa, 9 March.

_____ . (2002), *Food Security Strategy*, Addis Ababa: Federal Democratic Republic of Ethiopia.

Foucher, Abba Emile (1987), 'Birbirsa, 1868-1869' *Proceedings of the International Ssymposium on the Centenary of Addis Ababa*, November 24-25, 1986, Ahmed Zekaria, Bahru Zewde and Taddese Beyene (eds.), Addis Ababa: Institute of Ethiopian Studies, Addis Ababa University: 33-39.

Frey, Frederick W. (1993), 'The Political Context of Conflict and Cooperation over International River Basins,' *Water international*, 18: 54-68.

Gabre, Solomon (1994), 'Urban Land Issues and Policies in Ethiopia,' in *Land Tenure and Land Policy in Ethiopia After the Derg*, Proceedings of the Second Workshop of the Land Tenure Project, Dessalegn Rahmato (ed.), Addis Ababa, Institute of Development Research: 278-302.

Gebeyehu, Yibeltal (1996), 'Urban Environmental Problems, Planning and Management: The Case of Addis Ababa,' Paper Presented for the Workshop on *Urban-regional Development Planning and Implementation*, Organized by the National Urban Planning Institute and the International Development Association, 7-10 February.

Gebre, Betemariam, Ayele, Fekadu, and Shiferaw, Mahdere (1999), 'Patterns and Determinants of Under Five Mortality in Jimma Town, Jimma Zone, South West Ethiopia,' *Ethiopian Journal of Health Sciences*, 9, 2 (July): 101-108.

Gebre Egziabher, Tegegne, (1997) 'Decentralized Urbanization: An Urban Development Strategy for Ethiopia,' in *Urban and Regional Development Planning and Implementation in Ethiopia*, Tegegne G. Egziabher and Daniel Solomon (eds.), Addis Ababa: National Urban Planning Institute: 91-113.

Gebre Egziabher, Tegegne, (1998) 'Decentralization and Changing Local and Regional Development Planning,' in *Ethiopia in Broader Perspective*, Volume II, Papers of the 13th International Conference of the Ethiopian Studies, Katsuyoshi Fukui, Eisei Kurimoto and Masayoshi Shigeta (eds.): 691-720.

_____ . (2000), 'Housing Situation in Addis Ababa City: Constraints and Some Suggestions to Improve the Performance of the Sector,' paper presented at the symposium on *Building a Vibrant and a Livable Addis Ababa*, Addis Ababa Chamber of Commerce, 17-18 February.

Gleick, Peter, Wolf, Gary, Chalecki, Elizabeth L., and Reyes, Rachel, 'The Privatization of Water and Water Systems', in *The World's Water: The Biennial Report on Freshwater Resources* (Peter Gleick, ed.), Washington, D.C.: Island Press, 2002: 58-85.

Glennie, Colin (1983), *Village Water Supply in the Decade: Lessons from Field Experience*, New York: Wiley.

Gourlay, K. A. (1992), *World of Waste: Dilemmas of Industrial Development*, London: Zed Press.

GTZ-Urban Management Advisory Services (2000), *Addis Ababa Current Issues: Urban Development, Land Management, Formal and Informal Investment*, Addis Ababa.

Habte Mariam, Dejene (1987), 'Architecture in Addis Ababa,' in *Proceedings of the International Symposium on the Centenary of Addis Ababa*, November 24-25, 1986, Ahmed Zekaria, Bahru Zewde and Taddese Beyene (eds.), Addis Ababa: Institute of Ethiopian Studies, Addis Ababa University: 199-215.

Hailemariam, Assefa and Kloos, Helmut (1993), 'Population,' in *The Ecology of Health and Disease in Ethiopia*, Helmut Kloos and Zein Ahmed Zein (eds.), Boulder: Westview Press: 47-66.

Hardoy, Jorge E. and Satterthwaite, David (1992), *Environmental Problems in Third World Cities*, London: Earthscan Publications.

Hardoy, Jorge E., Mitlin, Diana, and Satterthwaite, David (2001), *Environmental Problems in an Urbanizing World*, London: Earthscan Publications.

Harpham, Trudy and Grant, Emma (2002), 'Health, Health Services and Environmental Health,' in *Urban Livelihoods: A People Centered Approach to Reducing Poverty*, Carole Rakodi and Tony Lloyd-Jones (eds.), London: Earthscan: 165-179.

Hartmann, Betsy (1997), 'Women, Population and the Environment: Whose Consensus? Whose Empowerment?' in *The Women, Gender and Development Reader*, Nalini Visvanathan, Lynn Duggan, Laurie Nisonoff, and Nan Wiegersma (eds.), London: Zed Press: 293-309.

Holman, C. (1995), 'Control of Pollutant Emissions From Road Traffic,' in *Pollution: Causes, Effects and Control*, Roy Harrison (ed.), London: The Royal Society of Chemistry: 293-317.

Holmes, Joan (2001), 'A Crisis of Leadership,' *The Hunger Project*, November. Horvath, Ronald J. (1968), 'Towns and in Ethiopia,' *Erkunde*, 22: 42-48.

Horvath, Ronald J. (1968), 'Von Thunen's Isolated State and the Area Around Addis Ababa, Ethiopia,' Department of Geography, East Lansing, Michigan University: 308-323.

_____ . (1968), 'Addis Ababa's Eucalyptus forest,' *Journal of Ethiopian Studies*, 6: 13-19.

_____ . (1969), 'The Wandering of Capitals of Ethiopia,' *Journal of African History*,10: 77-88.

Integrated Service for AIDS Prevention and Support Organization (ISAPSO) (2001), *Facts About HIV/AIDS*, Addis Ababa: United Nations Staff HIV/AIDS Workplace Education and Care Program.

Jacobs, G. D. (1997), 'Road safety in developing world,' in T. Fletcher A.J. Michaels (eds.) *Health at the Crossroads: Transport Policy and Urban Health*, Sage: London School of Hygiene and Tropical Medicine.

James A. and McDougal, F. R. (1995), 'The Treatment of Toxic Wastes,' in Pollution: Causes, Effects and Control, Roy M. Harrison (ed.), Cambridge: The Royal Society of Chemistry: 123-143.

Jehl, Douglas (1997), 'Nile-in-Miniature Tests Its Parent's Bounty,' *The New York Times*, 10 January.

Jira, Challi (1999), 'Reason for Defaulting from Expanded Program of Immunization (EPI) in Jimma Town, South Western Ethiopia,' *Ethiopian Journal of Health Sciences*, 9, 2 (July): 93-100.

Johansson, Richard (1996), *Survey of Liquid Waste and Sludge Treatment in Addis Ababa, Ethiopia*, Gotebor: Institutionen for vattenforsorjnings-och avaloppsteknik. Jovanovic, D. (1985), 'Ethiopian Interests in the Division of the Nile River Waters,' *Water International*, 10: 82-85.

Jowsey, Ernie and Kellett, Jonathan (1995), 'Sustainability and Methodologies of Environmental Assessment for Cities,' in *Sustainability, the Environment, and Urbanization*, Cedric Pugh (ed.), London: Earthscan: 197-227.

Kassu, A. and Aseffa, A (1999), 'Laboratory Service in Health Centers of Amhara Region in Northern Ethiopia,' *East African Medical Journal*, 76, 5: 239-242.

Kauzeni, A. S. (1983) 'Strategies and Factors Affecting Community Mobilization and Participation in Rural Water Development: A Case of Twelve Villages in Rukwa Region of Tanzania,' Paper presented on *Water for All: Coordination, Education and Participation*, held at the University of Linkiping, Sweden, May 29-June 4.

Kebbede, Girma (1992), *The State and Development in Ethiopia*, London: Humanities Press.

Ketema, Atsede (2000), 'Addis Ababa Plastic Bag Campaign and Recycling Urban Waste,' *Addis Tribune*, 21 April.

Khodakevich, Lev and Zewdie, Debrework (1993), 'AIDS' in *The Ecology of Health and Disease in Ethiopia*, Helmut Kloos and Zein Ahmed Zein (eds.), Boulder: Westview: 319-337.

Kidanu, Aklilu (2000), 'Why Fertility will Remain High in Ethiopia,' in *Proceedings of the Inaugural Workshop of the Forum for Social Studies*, 18 September 1998, Addis Ababa: Forum for Social Studies.

Kidanu, Aklilu and Banteyerga, Hailom (2003), 'HIV/AIDS and Poverty in Ethiopia,' Addis Ababa: Miz-Hasab Research Center, January.

Kidanu, Aklilu and Rahmato, Dessalegn (2000), *Listening to the Poor*, Forum for Social Studies Discussion Paper No. 3, Forum for Social Studies: 80- 99.

Kjellen, Marianne and McGranahan, Gordon (1997) Comprehensive Assessment of the Fresh Water Resources of the World: Urban Water, Towards    Health and Sustainability, Stockholm: Stockholm Environmental Institute.

Klare, Michael T. (2001), *Resource Wars: The New Landscape of Global Conflict*, New York: Metropolitan Books. Kossove, D. (1982), 'Smoke-Filled Rooms and Respiratory Disease in Infants,' '*South African Medical Journal*, 63: 622-624.

Kloos, Helmut and Tesfa-Yoyhannes, Tesfa-Michael (1993), 'Intestinal Parasitism,' in *The Ecology of Health and Disease in Ethiopia*, Helmut Kloos and Zein Ahmed Zein (eds.), Boulder: Westview Press: 223-235.

Kloos, Helmut and Zein, Zein Ahmed (eds.) (1993), 'Modern Health Services,' in their *The Ecology of Health and Disease in Ethiopia*, Boulder: Westview Press: 137-156.

Koren, Herman (1991), *Handbook of Environmental Health Safety: Principles and Practices*, Volume II, New York: Lewis Publishers.

_____ . (1996), *Illustrated Dictionary of Environmental Health and Occupational Safety*, CRC/Lewis Publishers.

Kristof, Nicholas D.(2003), 'Alone and Ashamed,' *The New York Times*, 16 May 20, Op-Ed.

Kumie, Abera (1997), 'An Overview of an Environmental Health Inspection Service in Addis Ababa,' in Region 14 Administration, Environmental Health Department, *A Comprehensive Overview of Addis Ababa Municipality Solid Waste Management and its Environmental Health Inspection Services*, Addis Ababa: Environmental Health Department, Region 14 Administration. 100-133.

Kusner, David J. (1993) 'Tuberculosis,' in Tropical and Geographical Medicine, Adel A.F. Mahmoud (ed.), New York: McGraw-Hill: 280-285.

Larkin, Maureen (1998), "Global aspects of health and health policy in Third World countries," in *Globalization and the Third World* Ray Kiely and Phil Marfleet (eds.), New York: Routledge, 1998: 91-111.

Larson, Charles P., MD and Desie, Tadele, MD (1994), 'Health in Ethiopia: A Summary of 52 District Health Profiles,' *The Ethiopian Journal of Health and Development*, 8, 2: 87-96.

Lidetu, Shewatatek and Okubagzhi, Gebreselassie (1993), 'Childhood Diseases and Immunization,' in *The Ecology of Health and Diseases in Ethiopia*, Kloos and Zein (eds.), Boulder: Westview Press: 191-201.

Ling, Ooi Geok (1999), 'Civil Society and the Urban Environment,' in *Cities and the Environment: New Approaches for Eco-Societies*, Takashi Inoguchi, Edward Newman, and Glen Paoletto (eds.), New York: United Nations University Press: 105-126.

Lo, C. T., Kloos, Helmut and Birrie, Hailu (1988), 'Schistosomiasis,' in *The Ecology of Health and Disease in Ethiopia*, Zein Ahmed Zein and Helmut Kloos (eds.), Addis Ababa: Ministry of Health.

Lopez, Murray C. (ed.) (1996) 'Global burden of disease and injury' In *Global Health Statistics*, Series Vol. 2, Cambridge, MA: Harvard School of Public Health.

Luoma, Jon (2002) 'Contamination, riots, rate of increases, scandals: From Atlanta to Manila, cities are confronting the true cost of water privatization,' *MotherJones*, November and December 2002: 35-38 and 88.

Mageed, Yahia Ablel (1994), 'The Nile Basin, Lessons from the Past,' *International Waters of the Middle East*, Asit K. Biswas (ed.), Bombay: Oxford University Press.

Malifu, Ammanuel (2000), 'Problems of the Urban Environment,' in *Proceeding of the Symposium of the Forum for Social Studies*, 15-16 September, Addis Ababa, Forum for Social Studies: 133-156.

Mara, Duncan (1996), *Low-Cost Urban Sanitation*, New York: Wiley.

Maurizio, Arcari (1997), 'The Draft Article on the Law of International Watercourses Adopted by the International Law Commission: An Overview and Some Recommendation on Selected Issues,' *Natural Resources Forum*, 21, 3: 169-179.

Mavalankar, D. V., Trivedi, C. R., Gay, R. H. (1999) 'Levels and Risk Factors for Perinatal Mortality in Ahmedabad, India,' *Bulletin of the World Health Organization*, 69, 4: 435-442.

McAuslan, P. (1994), 'Land in the City: The Rule of Law in Facilitating Access to Land by the Urban Poor,' Background Paper for the Global Report on Human Settlements, UNCHS.

McGranahan, Gordon (1991) *Environmental Problems and the Urban Household in Third World Countries*, Stockholm: The Stockholm Environment Institution.

McGranahan, Gordon, et. al. (1999), *Environmental Change and Human Health in Countries of Africa, the Caribbean and the Pacific*, Stockholm: The Stockholm Environmental Institute.

Meade, Melinda S. and Erickson, Robert J. (2000), *Medical Geography*, New York: The Guilford Press.

Megerssa, Negussie (2001), 'Report on Traveling Workshop on Obsolete Pesticides and Integrated Pest Management Alternatives,' *Solution*, 2, 9 (August): 2 and 6-7.

Mehrotra, S and Jolly, C. (eds.) (2000) *Development with a Human Face: Experiences in Social Achievement and Economic Growth*, Oxford: Oxford University Press.

Miller, G. Tyler (1998), *Living in the Environment*, Belmont: Wadsworth Publishing.

Ministry of Economic Development and Cooperation (MEDaC) (1999), *Survey of the Ethiopian Economy*, Addis Ababa: MEDaC.

Ministry of Health (1996a), *Study of Liquid Waste Management in the Urban Centers of Ethiopia*, Addis Ababa: Hygiene and Environmental Health Department, Ministry of Health.

_____ . (1996b), *Study of Water Quality Standard*, Addis Ababa: Hygiene and Environmental Health Department, MOH. Ministry of Health (1998a), *Health and Health Related Indicators*, Addis Ababa: Ministry of Health.

_____ . (1998b), *AIDS in Ethiopia: Background Projects, Impacts and Intervention*, Ministry of Health, Epidemiology and AIDS Department.

_____ . (1998c), *Hygiene and Environmental Health Policy*, Addis Ababa: Hygiene and Environmental Health Department, MOH.

_____ . (2000a), *Health and Health Related Indicators 1999/2000*, Addis Ababa: Ministry of Health.

_____ . (2000b), *Ethiopia's Health Care Financing Experience: Report on the Current Status of the Implementation of the Government's Health Care Financing Strategy*, Addis Ababa, Essential Services for Health in Ethiopia, MOH.

_____ . (2000c), *AIDS in Ethiopia*, Addis Ababa: Ministry of Health. Ministry of Health (2001), *Health and Health Related Indicators 2000/2001*, Addis Ababa: MOH.

_____ . (2002), *Health Care System Tier System With their Basic Parameters*, Addis Ababa: Health Information Team Office, MOH.

Ministry of Labor and Social Affairs (1991), *Survey of Adolescent Fertility, Reproductive Behavior and Employment Status of the Youth Population in Urban Ethiopia*, Report 1, Manpower and Employment Department, Addis Ababa: Ministry of Labor and Social Affairs.

Ministry of Public Works (1967), *Housing Study, Volume 1*, Ministry of Public Work, Imperial Ethiopian Government.

Ministry of Urban Development and Housing (1983), *Report on the Workshop on Basic Urban Services*, Addis Ababa, Ministry of Urban Development and Housing.

Ministry of Water Resources (1995), *Letter of Sector Policy*, Addis Ababa: Ministry of Resources.

_____ . (1996), *Tekeze River Basin Integrated Development Master Plan Project: Volume I, Main Report*, Addis Ababa: Ministry of Water Resources.

_____ . (1997a), *Water Supply Development and Rehabilitation: Technical Management Study*, Addis Ababa: MOWR.

_____ . (1997b), *Abbay River Basin Integrated Development Master Plan Project, Volume VII: Infrastructure*, Addis Ababa: MOWR.

_____ . (1997c), *Abbay river Basin Integrated Development Master Plan Project, Volume V: environment*, Addis Ababa: MOWR.

_____ . (1997d), 'Ministry of Water Resources and Its Current Major Activities,' *Water and Development*, Volume 2, 5: 20-29.

_____ . (1997e), *Water Supply Development and Rehabilitation: Technical Management Study*, Addis Ababa: MOWR.

_____ . (1998a), *Comprehensive and Integrated Water Resources Management*, Volume 1, Addis Ababa: MOWR.

_____ . (1998b), *Comprehensive and Integrated Water Resources Management*, Volume 2, Addis Ababa: MOWR.

_____ . (1998c), *Comprehensive and Integrated Water Resources Management*, Volume 3, Addis Ababa: Federal Water Policy and Strategy Development Project Office.

_____ . (1998d), *Water Supply Development and Rehabilitation: Tariff and Asset Valuation Study*, Addis Ababa: MOWR.

_____ . (2001a), *National Water Supply and Sanitation Master Plan, Status Report, Volume II: Physical Resource Base*, Addis Ababa: Ministry of Water Resources, Annex II-4.1.

_____ . (2001b), *National Water Supply and Sanitation Master Plan, Volume II: National Master Plan*, Addis Ababa: MOWR.

_____ . ( 2001c), *National Water Supply and Sanitation Master Plan, Status Report, Volume III, Financial and Economic Base*, Addis Ababa: MOWR.

_____ . (2002a), *Water Sector Development Program, Main Report, Volume II*, Addis Ababa: Ministry of Water Resources.

_____ . (2002b), *Water Sector Development Program: Investment Requirement Summary*, Addis Ababa: MOWR.

Ministry of Water Resources/Ernst &Young (1997a), *Water Supply Development and Rehabilitation Project: Socio-economic Baseline Studies, Volume I, Main Report*, Addis Ababa: MOWR.

_____ . (1997b), *Data Compilation and Analysis Project: Final Report*, Addis Ababa: MOWR.

_____ . (1998), *Water Supply Development and Rehabilitation: Tariff and Asset Valuation Study: Main Report*, Addis Ababa: MOWR.

Ministry of Works and Urban Development (1997), *Ethiopian Housing Sector Survey*, Addis Ababa: Ministry of Works and Urban Development.

_____ . (1999), *Legal Status, Role, Responsibilities and Relationships of Municipalities in the Amhara National Regional State*, Addis Ababa: Bureau of Works and Urban Development.

Moeller, Dade W. (1997), *Environmental Health*, Cambridge: Harvard University Press.

Moore, Gary S. (1999), *Living with the Earth: Concepts in Environmental Health Science*, New York: Lewis Publishers.

National Committee on Traditional Practices of Ethiopia (1999), *Major Harmful Traditional Practices in Ethiopia*, Addis Ababa: NCTPE.

Nwokeabia, Hilary (2002), *Why Industrial Revolution Missed Africa: A 'Traditional Knowledge' Perspective*, Addis Ababa: Economic Commission for Africa, ECA/ESPD/WPS/01/02.

Nyambura, Betty (2003), 'Privatization Not a Cure for Kenya's Water Woes,' *The East African*, 2 June.

Ocama, Ponsiano and Lee, William M. (2003), 'Don't Forget This Infectious Killer,' *The New York Times*, 1 March.

O'Dempsey, T. J. D., et. al. (1996) , 'A Study of Risk Factors for Pneumococcal Disease Among Children in a Rural Areas of West Africa,' *International Journal of Epidemiology*, 25, 4: 885-893.

Organization for Social Services for AIDS (1999), *Fight AIDS Together, Addis Ababa*: Addis Ababa: OSSA.

Palen, J. John (1981), *The Urban World*, New York: McGraw-Hill. Panafrican News Agency (2001), 'Maternity Mortality Still High in Africa,' 9 March.

Pankhurst, Richard (1962), 'The Foundation and Growth of Addis Ababa to 1935,' *Ethiopia Observer*, 6, 1: 33-61.

_____ . (1965), 'Notes on the Demographic History of Ethiopian Towns and Villages,' *Ethiopian Observer*, 9, 1: 60-83.

_____ . (1968), *Economic History of Ethiopia 1800-1935*, Addis Ababa: Addis Ababa University Press.

_____ . (1987), 'Development in Addis Ababa During the Italian Fascist Occupation,' in *Proceedings of the International Symposium on the Century of Addis Ababa*, 24-25 November 1986, Ahmed Zekaria, Bahru Zewde and Taddese Beyene (eds.), Addis Ababa: Institute of Ethiopian Studies, Addis Ababa University. 119-139.

_____ . (1990), An Introduction to the Medical History of Ethiopia, Trenton: The Red Sea Press.

Pankhurst, Richard and Ingrams, Leila(1988), *Ethiopia Engraved*, London: Kegan Paul International.

Pankhurst, Sylvia (1957), 'Sir Patrick Abercrombie's Town Plan,' *Ethiopian Observer* 1, 2: 34-44.

Pausewang, Siegfried, Tronvoll, Kjetil and Aalen, Lovise (eds.) (2002), *Ethiopia Since the Derg: A Decade of Democratic Pretension and Performance*, London: Zed Press.

Peters, Kim (1998), *Community-Based Waste Management for Environmental Management and Income Generation in Low-Income Areas: A Case Study of Nairobi, Kenya*, Toronto: Canada's Office of Urban Agriculture.

Phillipson, David W.(1994), 'The 1993 Excavations at Aksum,' in *New Trends in Ethiopian Studies: Papers of the Twelfth International Conference of Ethiopian Studies, Volume 1: Humanities and Human Resources*, Harold Marcus (ed.), Lawrenceville, N.J.: The Red Sea Press.

Platt, Anne (1998), 'Water-Borne Killers,' *Global Issues 1998/1999*: 58-64.

Polprasert, Chongrak (1996), *Organic Waste Recycling*, New York: Wiley. Policy and Human Resource Development Project (1998), *Ethiopian Social Sector Study Report: Cost-effectiveness of the Public Sector Health Care System*, Addis Ababa, Ministry of Health.

Postel, Sandra (2000), 'Redesigning Irrigated Agriculture,' *The State of the World 2000*, New York: Norton: 39-58.

Pruss, A., Giroult, E. and Rushbrook, P. (eds.) (1996), *Safe Management of* Wastes from Health-Care Activities, Geneva: World Health Organization.

Region 14 Administration (1997), *Regional Conservation Strategy, Volume I :Resource Base and Its Utilization*, Addis Ababa: Region 14 Administration.

Region 14 Administration Health Bureau (1997), *A Comprehensive Overview of Addis Ababa Municipality Solid Waste Management and its Environmental Health Inspection Services*, Addis Ababa: Environmental Health Department, Region 14 Administration. *Reporter* (2003), 'Solid Waste Management in Addis Ababa,' 23 February.

Rupphael, Yohannes (2000), 'Women Still Short-Changed in Ethiopian Society,' *Panafrican News Agency*, 11 May.

Sanders, Robert and Warford, Jeremy J. (1976), *Village Water Supply: Economics and Policy in the Developing World*, Baltimore: The Johns Hopkins University Press.

Satterwaite, David (2000), 'Lessons from the Experience of Some Urban Poverty-Reduction Programs,' in *Urban Livelihoods*, Carole Rakudi and Tung Lloyd Jones (eds.), London: Earthscan: 257-269. Save the Children (2001), *The Bitterest Pill of All: The Collapse of Africa's Health Systems*, London: Save the Children.

Shiferaw, Tesfaye (1999), 'Health Before 2000 in Ethiopia: Development, Achievements and Challenges,' in *Proceedings of the 10<sup>th</sup> Annual Scientific Conference of the Ethiopian Public Health Association, Health Before and After Year 2000 in Ethiopia*, 19-20 October.

Shiffman, Roy, (1997), 'Urban Poverty: The Global Phenomenon of Poverty and Social Marginalization in Our Cities: Facts and Strategies,' Keynote address at the 1995 Salzburg Congress on Urban Planning and Development Conference, New York.

Shinn, David (Ambassador) (2001), 'HIV/AIDS in Ethiopia: The Silence is Broken; The Stigma is Not.' *Africa Note*, No. 1, Washington, D.C.: Africa Program, Center for Strategic and International Studies.

Shiva, Vandana (1994), 'Women, Ecology and Health: Rebuilding Connections,' in *Close to Home: Women Reconnect Ecology, Health and Development Worldwide*, Vandana Shiva (ed.), Philadelphia: New Society Publishers: 1-42.

Simioni, Daniela (1999), 'An Institutional Capacity-Building Approach to Urban Environmental Governance in Medium-Sized Cities in Latin America and the Caribbean,' in *Cities and the Environment: New Approaches for Eco-Societies*, Takashi Inoguchi, Edward Newman, and Glen Paoletto (eds.), New York: United Nations University Press: 127-160.

Smith, Scott E. and Al-Rawahy, Hussain M.(1990), 'The Blue Nile: Potential for Conflict and Alternatives for Meeting Future Demands,' *Water International* 15: 217-222.

Speidel, J. Joseph (2002), 'Population, Consumption, and Human Health,' in *Life Support: The Environment and Human Health*, Michael McCally (ed.), Cambridge: MIT Press, 2002: 83-97.

Spellman, Frank R. and Drinan, Joanne (2000), *The Drinking Water Handbook*, Lancaster, PA: Technomic Publishing Co.

Stock, Robert (1995), *Africa South of the Sahara: A Geographical Interpretation*, New York: The Guilford Press.

Tadesse, Theodros, Bantayehu, Mesfin and G. Mariam, Lia (2000), *The Urban Transport and Road Network of Addis Ababa*, Addis Ababa: Addis Ababa Master Plan Revision Office.

Tagliabue , John (2002), 'As Multinationals Run the Taps, Anger Rises Over Water for Profit', *The New York Times*, 26 August.

Tchobanoglous, George and Burton, Franklin L. (1991), *Liquid Waste Engineering: Treatment, Disposal and Refuse*, New York: McGraw-Hill.

Tchobanoglous, George, Theisen, Hillary and Vigil, Samuel A. (1993), *Integrated Solid Waste Management: Engineering Principles and Management Issues*, New York: McGraw-Hill.

Teka, Gebre-Immanuel (1984), *Human Wastes Disposal in Ethiopia: A Practical Approach to Environmental Health*, Addis Ababa: The United Printers.

Tekle-Tsadik, Shimellis (1998), *Organizing the Housing Sector in the Principles of Free Market Economy*, Addis Ababa: Addis Ababa Chamber of Commerce.

Thomas, Leo and Taylor, Jon (2000) 'Squaring the Urban Circle: NGOs and Urban Poverty Alleviation in Addis Ababa,' *INTRAC Occasional Paper Series*, Number 24, February.

Tigabu, Tewodros (1991), *'Chereka Bet': A Unique Form of Illegal Housing Provision in Addis Ababa*, B.Sc. Thesis, College of Urban Planning, National Urban Planning Institute, Addis Ababa.

Todaro, Michael P. (1994), *Economic Development*, London: Longman. Transitional Government of Ethiopia (1995), *Health Sector Strategy*, Addis Ababa: MOH.

Tsige, Hailay (2001), 'Effect of Urban Solid Waste Compost on Soil Properties in Arba Minch, Ethiopia,' *Ethiopian Journal of Natural Resources*, 3, 1: 123-134.

Tulchin, Josehph S. (ed.) (1986), *Habitat, Health, and Development: A New Way of Looking at Cities in the Third World*, Boulder: Lynne Rienner Publishers. United Nations Program on HIV/AIDS (UNAIDS) (1999), *The UNAIDS Report*, Geneva: UNAIDS.

United Nations Program on HIV/AIDS (UNNAIDS) (2000a), *The HIV/AIDS Situation in Ethiopia*, Newsletter No. 4.

_____ . (2000b), *Report on the Global HIV/AIDS Epidemic*, Geneva: UNAIDS.

United Nations Center for Human Settlements (UNCHS) (1990), *The Global Strategy for Shelter to the Year 2000*, Nairobi: UNCHS.

_____ . (1996), *An Urbanizing World: Global Report on Human Settlements, 1996*, Oxford: Oxford University Press.

United Nations Development Program (UNDP) (1998), *Human Development Report 1998 1999*, New York: Oxford University Press.

_____ . (2000), *Human Development Report 2000*, New York: UNDP.

United Nations Development Program /World Bank (1997), *Environmental Sanitation: Case Study in Addis Ababa, Final Report*, Volume II, Addis Ababa: Region 14 Administration.

United Nations Economic Commission for Africa (1996), *Report on the State of Human Settlements in Africa*, Addis Ababa: Economic Commission for Africa.

United Nations Environmental Program (1992), *Report of a Workshop on Water Quality Monitoring and Assessment*, 23 March – 3 April, Arusha, Tanzania.

United Nations Children's Fund (1996), *The Protection, Rehabilitation and Prevention for Street Children and Street Mothers in Ethiopia*, Addis Ababa, UNICEF.

United Nations Children's Fund (1997), *The Progress of Nations: The Nations of the World Ranked According to their Achievement in Child Health, Nutrition, Education, Water, and Sanitation, and Progress for Women*, New York: United Nations.

United Nations Office for the Coordination of Humanitarian Affairs (2002), 'Ethiopia: Anti-AIDS Taskforce Criticized,' 6 November.

_____ . (2002), 'Ethiopia: Interview with Negatu Mereke, Head of the National AIDS Secretariat.'27 November.

United States Agency for International Development (2001),	*Ethiopia Integrated Strategic Plan*, Washington, D.C.: USAID, 24 September.

Vecchiato, Norbert L. 1993), 'Traditional Medicine,' in *The Ecology of Health and Disease in Ethiopia*, Helmut Kloos and Zein Ahmed Zein (eds.), Boulder: Westview Press: 157-178.

Vesilind, P. Aarne, Peirce, J. Jeffrey, and Weiner, Ruth F. (1990), *Environmental Pollution and Control*, Boston: Butterworth-Heinemann.

Waldman, Amy (2002), 'As AIDS Spread, India Struggles for a Workable Strategy,' *The New York Times*, 11 October.

Wallace, Helen M. (1995), 'Global View of Maternal and Child Health,' *Health Care of Women and Children in Developing Countries*, Helen M. Wallace, Kanit Giri, and Carlos V. Serrano (eds.), Oakland: Third Party Publishing Company: 12-38.

Walters, S. and Ayres, J. (1995) 'The Health Effects of Air pollution,' in *Pollution: Causes, Effects and Control*, Roy M. Harrison (ed.), Cambridge: The Royal Society of Chemistry.

Water and Sanitation for Health Project (1983), *Lessons Learned in Water, Sanitation and Health: Thirteen Years of Experience in Developing Countries*, Washington, D.C.: Water and Sanitation for Health, United States Agency for International Development.

Water Resources Development Authority (1995), *Survey and Analysis of the Upper Baro-Akobo Basin, Volume I, Main Report*, Addis Ababa: Water Resources Development Authority.

Water Supply and Sewerage Authority (1993), *Seven Towns Water Supply Project Environmental Impact Assessment Study*, Addis Ababa: Ministry of Water Resources.

Waterbury, John (2002) *The Nile Basin: National Determinants of Collective Action*, New Haven: Yale University Press.

Wekwete, Kadmiel H. (1997), 'Urban Management: The Recent Experience,' in *The Urban Challenge in Africa: Growth and Management of Its Large Cities*, Carole Rakodi (ed.), Tokyo: United Nations University Press: 527-552.

Whittington, Dale and McClelland, Elizabeth M. (1992), 'Opportunities for Regional and International Cooperation in the Nile Basin,' *Water International* 17: 144-154.

Wolde Mariam, Mesfin (1966), 'Problems of Urbanization in Ethiopia,' in the *Proceedings of the Third International Conference of Ethiopian Studies*, Addis Ababa: Institute of Ethiopian Studies, Addis Ababa University.

_____ . (1972), *Introductory Geography of Ethiopia*, Addis Ababa: Berhanena Selam Press.

Wolde Michael, Akalou (1967), *Urban Development in Ethiopia in Time and Space*, PhD Dissertation, University of California at Los Angeles.

World Bank (1994), *World Development Report 1994: Infrastructure for Development*, Washington, D.C.: The World Bank.

_____ . (1995), *Ethiopia: Water Supply Development and Rehabilitation Project*, Report No. 14401 ET, Addis Ababa: Ethiopia.

_____ . (1996), *Proceedings of the World Bank Sub-Saharan Africa Water Resources Technical Workshop*, February 12-15, Nairobi, Kenya.

_____ . (1998), *World Resources 1998-99*, New York: Oxford University Press.

_____ . (1999), *Intensifying Action Against HIV/AIDS in Africa: Responding to a Development Crisis*, Washington, D.C.

_____ . (2000a), *Cities Alliance for Cities Without Slums: Action Plan for Moving Slum Upgrading to Scale*, Washington, D.C.: The World Bank and UNCHS.

_____ . (2000b), *Human Development Report 2000*, New York: Oxford University Press.

_____ . (2000c), *World Development Report 2000/2001*, New York: Oxford University Press.

_____ . (2001a), *Ethiopia: Interim Poverty Reduction Strategy Paper 2000/2001-2001/2002-2003*, Report No. 21796-ET, 30 January.

_____ . (2001b), *World Development Report 2001/2002*, New York: Oxford University Press.

_____ . (2003), *World Development Indicators*, New York: Oxford University Press.

World Bank/United Nations Development Program (1992), *A Sanitation Sector Strategy for Ethiopia*, Addis Ababa.

World Health Organization (1948), *The Constitution of the World Health Organization*, Geneva: WHO.

_____ . (1972), *Health Hazard of the Human Environment*, Geneva: WHO.

_____ . (1992a), *Our Planet, Our Health*, Geneva: WHO.

_____ . (1992b), *Indoor Air Pollution from Biomass Fuel*, WHO/PEP/92.3A, Geneva: WHO.

_____ . (1996), *Mother-Baby Package: A List of Available Information*, Geneva: WHO.

_____ . (1998), *Female Genital Mutilation: An Overview*, Geneva: WHO.

_____ . (1999a), *Safe Management of Wastes from Health-Care Activities*, Geneva: WHO.

_____ . (1999b), *Guidelines for Air Quality*, Geneva: WHO.

_____ . (1999c), *Removing Obstacles to Human Development*, Geneva: WHO.

_____ . (2001a), *Maternal Mortality*, Geneva: WHO.

_____ . (2001b), *Water for Health: Taking Charge*, Geneva: WHO.

_____ . (2002), *Assessment and Planning Tool for Women's Health and Development: Indicators*, Addis Ababa: WHO.

World Health Organization/United Nations Children's Fund (1996), *Revised 1990 Estimates of Maternity Mortality: A New Approach*, Geneva: WHO.

World Health Organization/World Metrological Organization/United Nations Environmental Program (1996), *Climate Change and Human Health*, Geneva: WHO.

World Health Organization and United Nations Children's Fund (2000), *Global Water Supply and Sanitation Assessment Report*, Geneva: WHO.

World Resources Institute (1996), *The Urban Environment, 1996-97*, Oxford: Oxford University Press.

World Resources Institute (1998), *World Resources 1998-1999: A Guide to the Global Environment*, New York: Oxford University Press.

Wubneh, Mulatu and Yohannes (1988), *Tradition and Development in the Horn of Africa*, London: Avebury.

Yassi, Annalee, Kjellstrom, Tord, Kok, Theo de, and Tee L. Guidotti, Tee L. (2001), *Basic Environmental Health*, Oxofrd: Oxford University Press.

Zahr, C. Abou (1997), 'Improved Access to Quality Maternal Health Services, Paper Presented at the Safe Motherhood Technical Consultation in Sri Lanka, 18-23 October.

Zeitz, Paul S. (2003), 'Waging a Global Battle More Efficiently,' *The New York Times*, 1 March.

Zewde, Bahru (1987), 'Early Safars of Addis Ababa: Patterns and Evolution,' in *Proceedings of the International Symposium on the Centenary of Addis Ababa*, November 24-25, 1986, Ahmed Zekaria, Bahru Zewde and Taddese Beyene (eds.), Addis Ababa: Institute of Ethiopian Studies, Addis Ababa University: 43-55.

# Index

Aba Samuel, Lake 103, 114-15, 156
Abercrombie, Patrick Sir 29-30
Abijata-Shala National Park 103
Acute respiratory infections 2, 164-65, 203
Adama 2, 26, 32-33, 91-92, 108, 123, 163, 196
Addis Ababa
  as primate city 35
  development of 27-31
  employment/unemployment in 37-39
  flooding in 50, 113, 114, 124-25
  governance 60-63; see also Urban
  housing; see also Urban
    condition of 43, 48-51, 53
    informal settlements 51-53
    land lease system regulations in 44-47
    shortage 42, 50-60
  liquid waste management in 11-17; see also Liquid waste
  sanitation in 109-110
  solid waste management in 120-27, 130-31; see also Solid waste
  transportation 55-59; see also Urban
  unaccounted-for-water in 96
  water supply 85-90, 92
Addis Ababa University 178
Addis Ababa Water Supply and Sewerage Authority 89-90, 110-11
Addis Zemen 63
Adigrat 25
Adulis 24
Adwa 25, 28
Afar 5, 70, 97, 102, 14, 166, 182-83, 186-88, 190, 193-94, 203
African Development Bank 102, 115

Agriculture 9-12, 40, 80, 165, 178
Agriculture Development Led Industrialization 10-11
Ahmad Gragn 25
AIDS, see HIV/AIDS
Akaki 26-27, 31, 156
Aksumite Kingdom 24
Alaba Qulito 94, 99
Alemaya, Lake 94
Alemaya College of Agriculture/University 178
Amhara 5, 35, 90, 92, 97-98, 108, 123, 146-47, 163, 166, 182-84, 186-88, 190, 193-94, 203, 208-209
Amhara Regional State, 63-64, 184, 195; see also Amhara
Angola 185
Ankober 25
Arba Minch 32-33, 203
Asela 32-33, 41, 178, 208
Asmara 173
Asosa 92, 108, 123, 163
Assayita 92, 94, 99, 108, 123, 163
Australia 27
Awasa 2, 7, 32, 36-37, 92, 103, 108, 123, 146, 166, 178, 197
Awasa, Lake 103
Awash 26

Bahir Dar 2, 27, 32-33, 36-37, 41, 63-64, 69, 91-92, 109, 122-23, 146-47, 163, 196-97
Bale 102
Belgium 102
Benishangul-Gumuz 5, 90, 93, 97, 123, 166, 182-183, 187-88, 190, 193, 203
Biochemical oxygen demand (BOD) 113, 153-55, 157-58
Bishoftu 26, 32-33, 41, 59, 92, 95, 108, 163
Blue Nile 76-79

Burundi 79-80
Butajira 94, 99

Cameroon 58
Canada 102
Central African Republic 185
Chemical oxygen demand (COD) 153-56, 157-58
Composting 139-39
Construction and Business Bank (CBB) 47

Debre Berhan 25, 32
Debre Markos 25, 32
Debre Tabor 25, 95
Debub University 178
Deghabur 94-95, 99
Democratic Republic of Congo 75, 79-80, 185
Derg regime 4, 30, 41, 43-44, 174-75, 206
Dessie 32-33, 37, 41, 63-64, 91-92, 97, 108, 122-23, 163
Dilla 94-95, 99, 197
Dire Dawa 5, 26-27, 32-33, 35, 37, 41, 90-92, 94, 97, 122, 163, 166, 182-83, 186-88, 190, 193, 208-09
Dissolved oxygen (DO) 153, 157-58
Djibouti 26, 28, 77

Education sector 12-13
Egypt 24, 79-81
Environmental Protection Authority 132, 168, 226
Eritrea 4, 8, 24, 17, 19, 24, 37, 76, 80, 146, 174, 183
Ethiopian People's Revolutionary Democratic Front (EPRDF) 4, 7
Ethiopian Women Lawyers' Association 195
European Union 102
Expanded Program of Immunization 178

Family planning 23, 185
Famine 4, 102; *see also* Hunger
Fasil, Emperor 25
Federal constitution 5, 62

Female genital mutilation (FGM), *see* Women
Finfine 27
Finland 102
Fitche 96
Food and Agriculture Organization (FAO) 166
Food
    safety 117-18
    security/insecurity 9-12, 14, 192

Gambela 5, 90, 92-93, 95, 97, 108, 115, 123, 138, 146, 163, 182-83, 187, 190, 193, 197, 203
Gashe Abera Molla Environmental Movement 133-137
Germany 102
Ghana 58, 199
Gondar 25, 32-33, 63-64, 92, 108, 122-23, 163, 172, 174, 178

Haile Mariam, Mengistu Colonel 4
Haile Silassie, Emperor 86
Harar 25, 32-33, 35, 41, 93-95, 99; *see also* Harari
Harari 5, 25, 35, 90, 92, 93-94, 97, 107-108, 122-23, 146-47, 163, 182-83, 186-88, 190, 194; *see also* Harar
Hawzen 25
Healthcare
    development 172-74
    expenditures 12, 182-83
    facilities, 173-74, 177-79, 181-82, 184
    immunization 188-90
    insurance 209
    personnel 173-74, 179, 183
    policies 174-81
    reproductive 184-90
    user fees 208-09
    waste 127-31
    waivers/exemptions 207-08
Highly Indebted Poor Countries (HIPCs) 13-14
HIV/AIDS 3, 14, 23-24, 194-202
House of Federation 5
House of People's Representatives 5
Hunger 4; *see also* Famine

India 24, 162, 165, 195
Industries
distribution of 147
environmental effects of 153-62
types of 149-53
waste disposal methods of 151-53
International Drinking Water Supply
and Sanitation Decade 74-75
International Monetary Fund (IMF) 13-
14, 37, 203
Italy, Italian occupation 26-28, 86-87,
102, 166

Japan 102, 166
Jijiga 32,-33, 91-92, 108, 122-23, 163
Jimma 32-33, 37, 97, 108, 123, 163,
174, 178

Kaliti Sewage Treatment Plant 39, 113-
14, 116, 157
Kenya 58, 75, 79-80, 107
Kombolcha 63-64, 69, 146

Lalibela 24-25
Lalibela, King 24
Liberia 185
Little Akaki River 39, 86, 113-15, 156-
58
Liquid waste; *see also* Addis Ababa
health effects 116-18
management 110-16
Low-density polyethylene plastics
(LDPE) 124

Malaria 199, 202-04
Mali 185
Malnutrition 3, 12, 191-92
Mekele 2, 25, 32, 36-37, 41, 92, 107-
108, 122-23, 146-47, 163, 178,
208
Menelik II, Emperor 26-28, 172
Metahara 96
Millennium Development Goals
(MDGs) 14
Ministry of Health (MOH) 84, 100, 131-
32, 172, 181, 197, 206-7, 210, 222
Ministry of Trade and Industry 168
Ministry of Urban Development and
Housing (MUDH) 42

Ministry of Water Resources (MOWR)
75, 82-83, 100
Mohammed International Development,
Research, and Organization
Companies (MIDROC) 54-55, 59
Mojo 26
Mount Hakim 25, 143

Negele Arsi 96
Nekemte 7, 32-33, 178
Netherlands 102, 166
Niger 185
Nigeria 22, 120

Oromiya 5, 35, 90, 92-93, 97-98, 108,
123, 146, 163, 166, 182-83, 186,
188, 190, 193-94, 208-09
Oromiya Regional State, *see* Oromiya

Parliament 5, 7
Pesticides 10, 165-67
Pollution; *see also* Urban
sources of 148-53
effects of 153-58
air 159-63
indoor 162-65
Population
fertility 23-24, 178
life expectancy 3
rate of increase 22
family planning 23; *see also*
Women
Private sector/privatization 4, 10, 16, 37,
44, 49, 63, 84-85, 101, 111,131,
176, 179, 181-82, 200, 205, 220-
21, 223-24, 226-28

Qabri Dehar 94, 99

Railway, Addis Ababa-Djibouti 26, 28-
29
Red Sea 24
Ring Road 59

Sahle Silassie, King 27
Sanitation; *see also* Urban
state of 109-10
health effects of 138-42
Sasakawa Foundation 166

Shashamene 32-33, 41, 97, 197
Sierra Leone 185
Soddo 33, 197
Solid waste; *see also* Addis Ababa
    collection 120-24
    disposal 125-27
    health effects of 121-31
    recycling/minimization 126, 137-
      39
    scavenging 126
    sources of 118-20
Somali 5, 35, 90, 93, 146-47, 166, 182,
    186, 188, 190, 193-94, 203
Somali Regional State 25; *see also*
    Somali
Somalia 77, 79
South Africa 165, 195
Southern Nations, Nationalities and
    Peoples' Region (SNNPR) 5, 13,
    36, 90, 92-93, 97, 123, 146-47,
    163, 182-83, 186, 188, 190, 193-
    94, 203, 208-09
Sudan 76, 79-81
Sweden 102, 166

Taitu, Empress 27
Tanzania 79-80, 107
Tepi 7
Tertiary sector 9, 10
Tigray 5-6, 24, 36, 90, 93, 97, 147, 166,
    174, 182-83, 186-89, 193, 203,
    208-09
Tigray People's Liberation Front
    (TPLF) 4
Tigray Regional State 25; *see also*
    Tigray
Trade unions 4, 202
Traditional harmful practices (HTPs)
    194-95; *see also* Women
Traditional medical practice 206-207
Tuberculosis 2, 199, 202, 205

Uganda 75, 79-80, 107, 200-01
United Nations 60, 74, 102, 107, 203
United Nations Children's Fund
    (UNICEF) 74, 191-92
United Nations Development Program
    (UNDP) 102

United States Agency for International
    Development (USAID) 23, 166
Universal Declaration of Human Rights
    60, 171
Urban
    agriculture 39-40
    employment/unemployment 2, 11,
      14, 36-40
    governance 1, 60-69
    growth and development 24-31
    housing 40-54, 218-21
    informal sector 38-40
    informal settlements 51-53, 166
    land lease system 44-54
    population growth 31-36
    transportation 54-59

Wastewater; 111, 152, 223-24; *see
    also* Liquid waste
Water
    operation and maintenance 73-74,
      81, 84-85, 95-97, 100-102
    pollution 89, 93-94, 100, 103, 114-
      15
    quality 86, 89-90, 94-97, 99-
      100, 103
    related diseases 104-105
    resources 75-78
    sector development 82-85, 105
    shortage 86-89, 93-94, 105
    tariffs 83-85, 93, 96-99 105
    treatment 86-90, 96, 99, 104
    unaccounted-for 96, 99, 105
    vendors 93, 95, 97
Wolkite 94, 99
Women
    abuse 34-35, 191-95
    education 12-13
    employment/unemployment 34,
      37-38
    exposure to pollution 164-65
    HIV/AIDS 196-97
    malnutrition 191-92
    reproductive health 185-93; *see
      also* Healthcare
World Bank 8, 13-14, 37, 97, 203, 210,
    224

World Health Organization (WHO) 74-75, 84, 107, 127, 172, 174, 185, 192, 203, 210

World Trade Organization (WTO) 224

Yirgalem 178

Zara Yacob, Emperor 25
Zula 24

Printed and bound by CPI Group (UK) Ltd, Croydon, CR0 4YY

21/10/2024

01777109-0001